WITHDRAWN
FROM
UNIVERSITIES
AT
MEDWAY
LIBRARY

Robot sensors and transducers

Open University Press Robotics Series

Edited by

P.G. Davey CBE MA MIEE MBCS C.Eng

This series is designed to give undergraduate, graduate and practising engineers access to this fast developing field and provide an understanding of the essentials both of robot design and of the implementation of complete robot systems for CIM and FMS. Individual titles are oriented either towards industrial practice and current experience or towards those areas where research is actively advancing to bring new robot systems and capabilities into production.

The design and overall editorship of the series are due to Peter Davey, Managing Director of Meta Machines Limited, Abingdon; Fellow of St Cross College, Oxford University; and formerly Co-ordinator of the UK Science and Engineering Research Council's Programme in Industrial Robotics.

His wide ranging responsibilities and international involvement in robotics research and development endow the series with unusual quality and authority.

TITLES IN THE SERIES

Industrial Robot Applications	E. Appleton and D.J. Williams
Robotics: An Introduction	D. McCloy and M. Harris
Robots in Assembly	A. Redford and E. Lo
Robot Sensors and Transducers	S. R. Ruocco

Titles in preparation

Integration of Robots with Manufacturing Systems	R. Weston, C. Sumpter and J. Gascoigne
Unsolved Problems in Robotics	R. Popplestone

Robot sensors and transducers

S. R. Ruocco

HALSTED PRESS
John Wiley & Sons
New York – Toronto
and
OPEN UNIVERSITY PRESS
Milton Keynes

Open University Press
Open University Educational Enterprises Limited
12 Cofferidge Close
Stony Stratford
Milton Keynes MK11 1BY, England

First Published 1987

British Library Cataloguing in Publication Data

Ruocco, S.R.
Robot sensors and transducers.—(Open University robotics series).
1. Robots 2. Pattern recognition systems
I. Title
629.8′92 TJ211

ISBN 0-335-15410-7

ISBN 0-335-15408-5 Pbk

Published in the U.S.A., Canada and Latin America by
Halsted Press, a Division of John Wiley & Sons, Inc.,
New York.

Library of Congress Cataloging-in-Publication Data

Ruocco, S. R.
Robot sensors and transducers.

(Open University Press robotics series)
Bibliography: p.
Includes index.
1. Robots—Equipment and supplies. 2. Transducers.
3. Detectors. I. Title. II. Series.
TJ211.R86 1987 629.8′92 87-14838
ISBN 0-470-20894-5

Text design by Clarke Williams
Typeset and printed in Great Britain

Contents

Series Editor's Preface

The use of sensor's with machines, whether to control them continuously or to inspect and verify their operation, can be highly cost-effective in particular areas of industrial automation. Examples of such areas include sensing systems to monitor tool condition, force and torque sensing for robot assembly systems, vision-based automatic inspection, and tracking sensor's for robot arc welding and seam sealing. Many think these will be the basis of an important future industry.

So far, design of sensor systems to meet these needs has been (in the interest of cheapness) rather *ad hoc* and carefully tailored to the application both as to the transducer hardware and the associated processing software. There are now, however, encouraging signs of commonality emerging between different sensor application areas. For instance, many commercial vision systems and some tactile systems just emerging from research are able to use more or less standardized techniques for two-dimensional image processing and shape representation. Structured-light triangulation systems can be applied with relatively minor hardware and software variations to measure three-dimensional profiles of objects as diverse as individual soldered joints, body pressings, and weldments. Sensors make it possible for machines to recover 'sensibly' from errors, and standard software procedures such as expert systems can now be applied to facilitate this.

The growing industrial importance of sensor's brings with it the need to train (or re-train) many more engineers versatile enough to design, install and maintain sensor-based machine systems. The emergence of some commonality brings the opportunity to teach the subject in a systematic way. Accordingly, the author of this book explains the use of sensors both for 'internal' sensing—or encoding—of position and velocity within servo controlled robot axes, and for 'external' sensing to relate the robot to its task. A thorough and systematic presentation of the basic physical principles used in different single-cell devices leads the student to understand how the same principles underlie the more complex operation of linear and area array devices. The final section outlines standard methods of processing array data to produce reliable control signals for robot and other machines.

The book will have earned its place in the Open University Press Robotics Series if it achieves one thing: better training of robot engineers not just in what the sensing systems available to them do, but in how they work and thus how they may best be matched to future applications.

P. G. Davey

Preface

The requirements for the book were generated, initially, by the Middlesex Polytechnic decision to run on an Industrial Robotics course for the Manpower Services Commission based on the B. Tech. Industrial Robotics syllabus which includes approximately 25% of it on sensors. An added requirement for the book came from the need to integrate Robotics, rapidly becoming a subject area in its own right, into the curriculum of both the full time and part time Electronics degree courses.

The lack, at the time, of suitable books covering the required area of interest was a prime motive behind the decision to write extensive lecture notes which, after integration with the experience gained during the years of industrial employment, were edited into a format suitable for publishing.

The book is aimed at the student of an electronic discipline studying the traditional subjects of Systems, Control and Instrumentation as well as the more recently established one of Robotics. The level catered for is equivalent to the 2nd and 3rd year of a degree course, but since no prior knowledge of the principles involved is assumed the book is thought to be suitable for Diploma students as well.

A feature of this book is that it covers only the sensors and transducers necessary for computer interfacing as related to robot control. It therefore differs from similar books by omitting the sections on fluid-mechanical (density, flow, humidity and level), nuclear, temperature and chemical transducers and concentrating on the recent research and developments in the sensors and transducers suitable for interfacing to the 2nd and 3rd generation of robots. It also differs from other books on robot technology by omitting the sections on robot characteristics in favour of a section on Image Processing, thus providing a broad picture of modern sensing equipment and its applications in the modern field of Robotics.

Acknowledgements

The author wishes to thank all who have contributed to the compilation and the successful completion of this book, in particular my wife Patricia for her unfailing support and my Middlesex Polytechnic colleagues in the faculty of Engineering, Science and Mathematics for their constructive contributions.

The author wishes to credit the following sources for some of the figures, tables and other material used in this book; a reminder of each contribution is also included in the appropriate parts of the text: Academic Press Inc. (London) Ltd., 24/26 Oval Road, London NW1, UK; Apple User and Apple Computer Inc., 10260 Bandley Drive, Cupertino, CA, 95014, USA; Drews, S. P. *et al.*, Institut fur Prozesstenerung in der Schweissteck-nik, RWTH Achen, Rentershagweg 4, D-5100, Aachen, Federal Republic of Germany; Electronic Automation, Haworth House, 202 High Street, Hull, HU1 1HA, UK; Elsevier Science Publ. B.V., P.O. Box 1991, 1000 BZ, Amsterdam, The Netherlands; Fairchild Semiconductors Ltd., 230 High Street, Potters Bar, Herts, UK; Ferranti Electronics Ltd., Fields New Road, Chadderton, Oldham, OL9 8NP, UK; General Electric Semiconductors, Belgrave House, Basing view, Basingstoke, Hants, RG21 2YS, UK; Hewlett-Packard (UK) Ltd., Nine Mile Ride, Easthampstead, Wokingham, Berks, RG11 3LL, UK; Honeywell-Visitronics, P.O. Box 5077, Englewood, CO, USA; Joyce-Loebl, Marquis way, Team valley, Gateshead, Tyne and Wear, NE11 OQW, UK; Kanade, T. and Somner, T., Carnegie-Mellon University, Pittsburgh, PENN USA; PERA (Production Eng. Res. Assoc.), Melton Mowbray, Leicestershire, LE13 OPB, UK; Polaroid (UK) Ltd., Ashley Road, St. Albans, Herts, AL1 5PR, UK; Ramakant Nevatia, 'Machine perception', © 1982, pp. 159–163, Adapted by permission of Prentice-Hall Inc., Englewood Cliffs, NJ, USA; Robertson, B. and Walk-den, A., GEC-Hirst Research Labs., East End Lane, Wembley, Middlesex, UK; Rosenfeld, A. *et al.*, University of Maryland, College Park, Washin-gton D.C., USA; SPIE, The Int. Soc. of Opt. Engineers, P.O. Box 10, Bellingham, Washington, USA; Texas Instruments Ltd., Manton Lane, Bedford, MK41 7PA, UK; West, G. A. W. and Hill, W. J., City University, Northampton Square, London EC1, UK.

Chapter 1

Introduction

1.1 Historical notes

Man's perception of his environment has traditionally been limited by his five senses. As he tried to gain wider control of the environment by delving deeper into its physical nature, for instance through microscopy and astronomy, he reached the limitations of his human senses and therefore generated the need for devices that could help him in his continuing quest.

Since these devices were meant originally as an extention of his human senses it followed that they should be called *sensors*. In answer to the requirement to 'see' beyond the human colour spectrum, for instance, there came the infrared sensor and in response to the need to perceive sounds higher than the human ear could cope with, there came the ultrasonic sensor.

In the majority of cases these tasks were accomplished by transforming the physical input energy into an electrical output signal that could be measured and/or displayed on a suitable device. Sensors came, therefore, to be known as 'energy transformers', devices capable of converting a physical input quantity into an electrical output quantity. This process has been identified in various disciplines as a process of 'energy transduction' which has also led to sensors' other common name, *transducers*.

In fact, the differences in nomenclature go even further—as well as the aforementioned historical reasons there are also linguistic–geographical ones (there are differences in the terminology used in some European countries as well as some differences between USA and British terminology) and many within the various fields of physics too. Transducers are sometimes referred to as 'transmitters' in the process industry—e.g. pressure transmitter, as 'detectors' in the optical field—e.g. light detector, or as 'probes' in fluid–mechanical measurements—e.g. pressure probe. It is beyond the scope of this book to propose a rationalization of the existing nomenclature but definitions of what is intended here by sensor and transducer are required both to help the reader and to permit comparison with other books in this subject.

1.2 Definitions of sensor and transducer

A *transducer* is here defined as an elementary device capable, within a given field of measurement, of converting a physical non-electrical input quantity (the measurand) into an electrical output quantity. The transducer itself does not contain any further processing beyond this energy conversion.

A *sensor* is here defined as a non-elementary device, usually based on a transducer, capable of converting a physical non-electrical input quantity into an electrical output quantity *and* of processing it, in accordance with a given algorithm, to provide an output suitable for interfacing to a process control system such as a computer.

The main difference lies therefore in the non-elementary nature of the sensor, that is in its capability to embody functions other than the basic energy conversion. This leads to a further classification of sensors, that of:

- intelligent sensors, those that can interact with the control computer to provide data manipulation (such as 'feature extraction' as in the case of vision sensors)
- non-intelligent sensors, those that can provide the computer with the output data only (and which therefore require longer communication times).

1.3 Generalities

Both sensors and transducers can be classified according to their input/output characteristics. With respect to their input physical quantity these devices can be termed:

Absolute—when, given a fixed origin, the electrical output signal can represent all the possible values of the input physical signal with no ambiguity

Incremental—when an origin cannot be fixed for all points within the field of measurement and each point is taken as the origin for the next one

The nature of the output function, on the other hand, determines whether the device is:

Analogue—when the output signal is continuous and proportional to the input physical quantity

Digital—when, given a continuous physical input quantity, the output signal can only take a number of discrete values

The nature of the electrical output quantity also determines the device performance characteristics, which can be divided into *static* and *dynamic* ones.

Static characteristics describe the device performance at room conditions (i.e. at a temperature of 25 ± 10 degrees Celsius, a relative humidity of 90% or less and a barometric pressure of 880–1080 mbar), with very slow changes in the measurand and in the absence of any mechanical shock (unless this latter happens to be the measurand). Static characteristics include: linearity, accuracy, stability, precision, sensitivity and resolution.

Linearity—can be defined as the variation in the constant of proportionality between the input physical quantity and the output electrical signal. A sensor or a transducer is therefore said to be linear when the constant of proportionality has the same value within the whole measurand range (i.e. when the graph relating input to output is a straight line). The linearity error is expressed as a percentage of the maximum electrical output value.

For example, in Figure 1.1 the dotted line shows the theoretical straight line of maximum linearity within the measurand range. Curve 'a' shows the output characteristic for a device which is linear up to 70% of the measurand range and then exibits increasing deviation from straight line linearity. This behaviour is often known as 'output saturation' and affects most transducers, particularly optical ones.

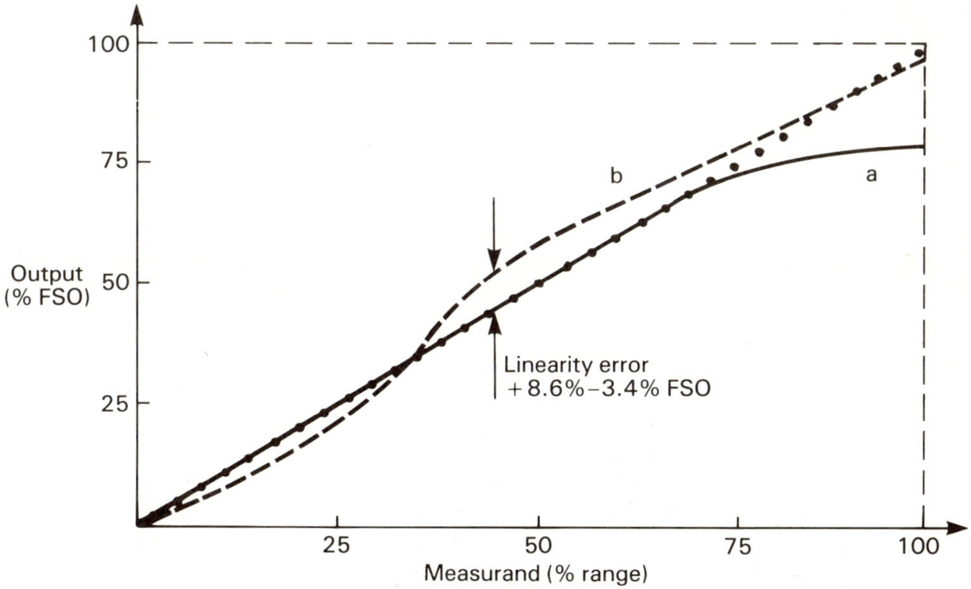

Figure 1.1

Curve 'b' shows the output characteristic of a transducer with a marked non-linear behaviour throughout the measurand range. The linearity error for device 'b' would therefore be +8.6%, −3.4% FSO (full-scale output) *with respect to the terminal linearity curve.* In fact it is meaningless to quote a linearity error without stating what sort of straight line it uses as reference. The choice of this straight line depends on the device output curve and its applications; as shown in Figures 1.2–1.5 there are four main types of curve used for linearity calculations:

(a) *Theoretical slope*
The straight line between the theoretical end points. Usually 0–100% for both FSO and range—in which case it is called the 'terminal line' (see dotted line in Figure 1.2)—or it can purposely be offset to suit certain applications (see broken line).

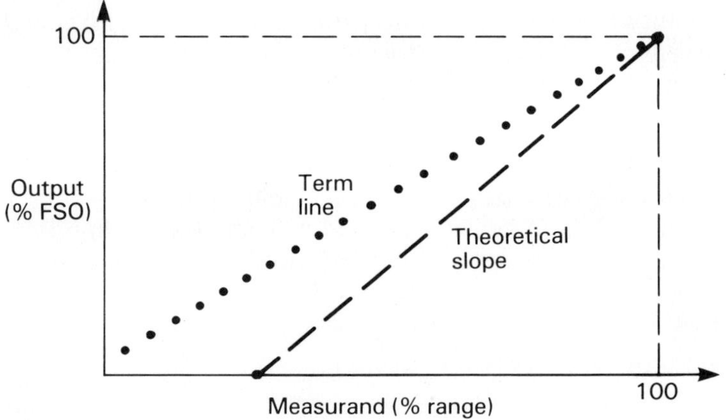

Figure 1.2

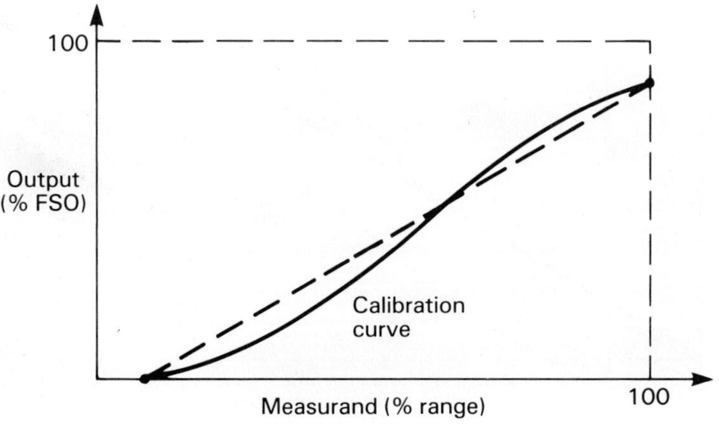

Figure 1.3

(b) *End-point line*
The straight line between the end points, namely the output values at the upper and lower measurand range limits obtained during any one calibration (see broken line in Figure 1.3).

(c) *Best straight line*
A line midway between the two parallel straight lines which are closest together and envelop all the output values obtained during calibration (see broken line in Figure 1.4).

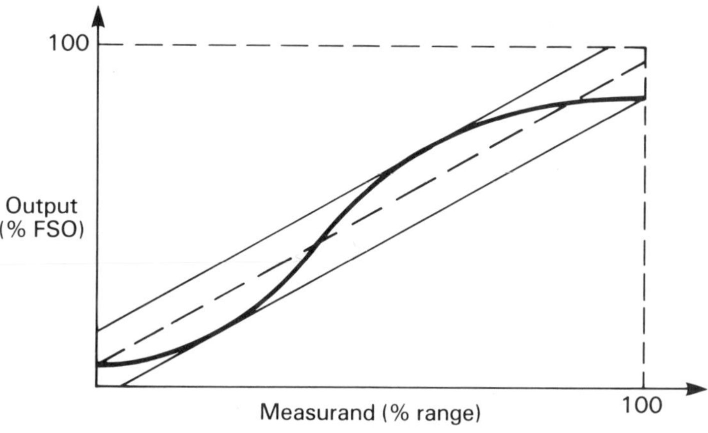

Figure 1.4

Precision—can be defined as the tolerance within which a measurement can be repeated (e.g. the ability of a device to give the same output value

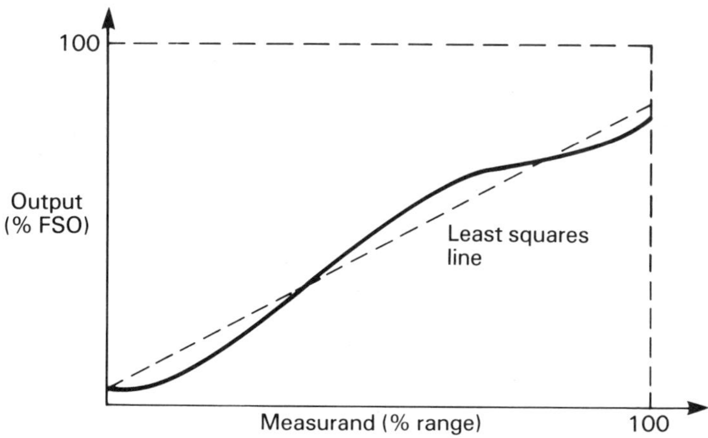

Figure 1.5

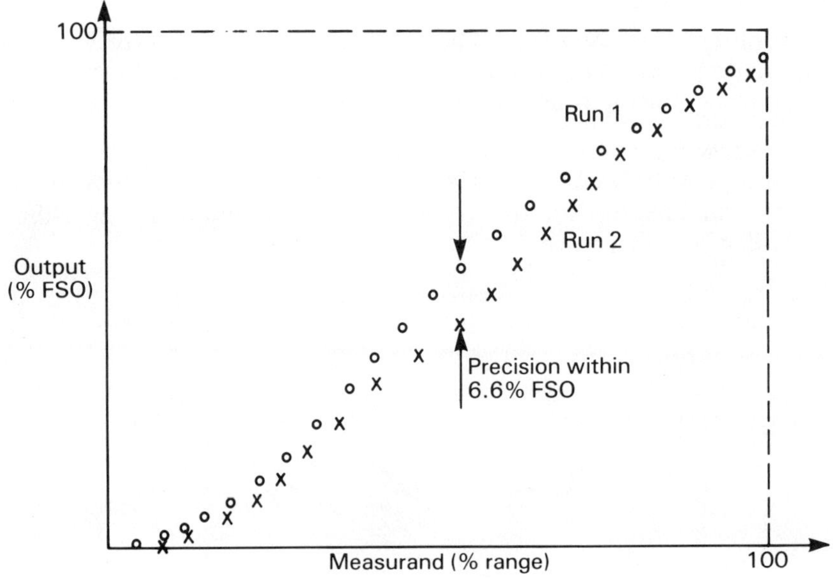

Figure 1.6

when the same input value is applied to it). Precision is normally expressed as a percentage of FSO and is sometimes referred to as *repeatability*. Note that the measurement must be carried out always in the same way to avoid hysteresis problems. See Figure 1.6.

Hysteresis—can be defined as the maximum difference in the output values

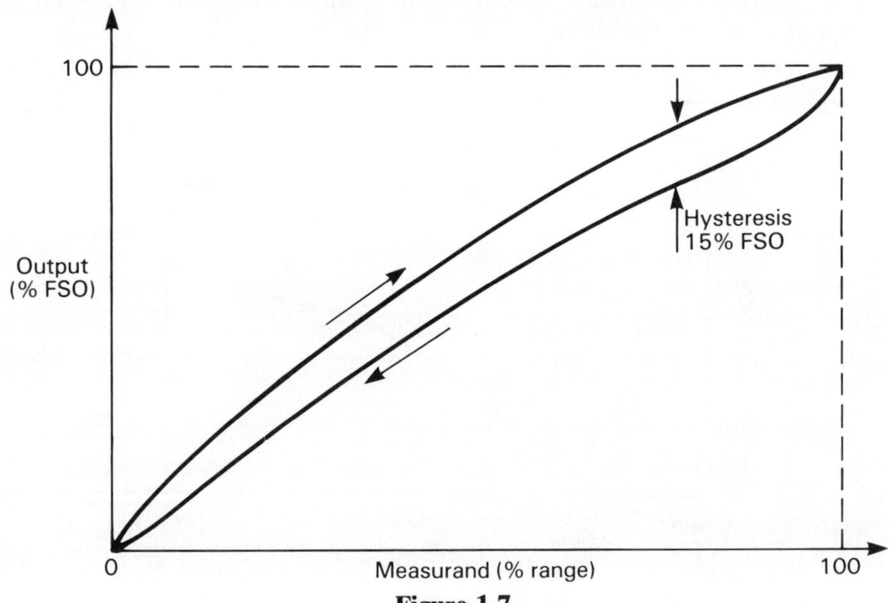

Figure 1.7

obtained by covering the measurand range first in the increasing direction (i.e. from zero to 100%) and then in the decreasing direction (i.e. from 100% to zero). See Figure 1.7.

Sensitivity—can be defined as the relationship between the variations in the output electrical quantity and the corresponding variations in the input physical quantity.

Resolution—can be defined as the magnitude of the output step changes produced by a continuously varying measurand. Normally expressed as a percentage of FSO. For digital output devices resolution is given by the number of bits in the output data word(s) or, in the case of incremental position transducers, by the number of states obtained per unit measurand change. Figure 1.8 shows the output for a wirewound potentiometer with a very poor resolution of 2.5% FSO.

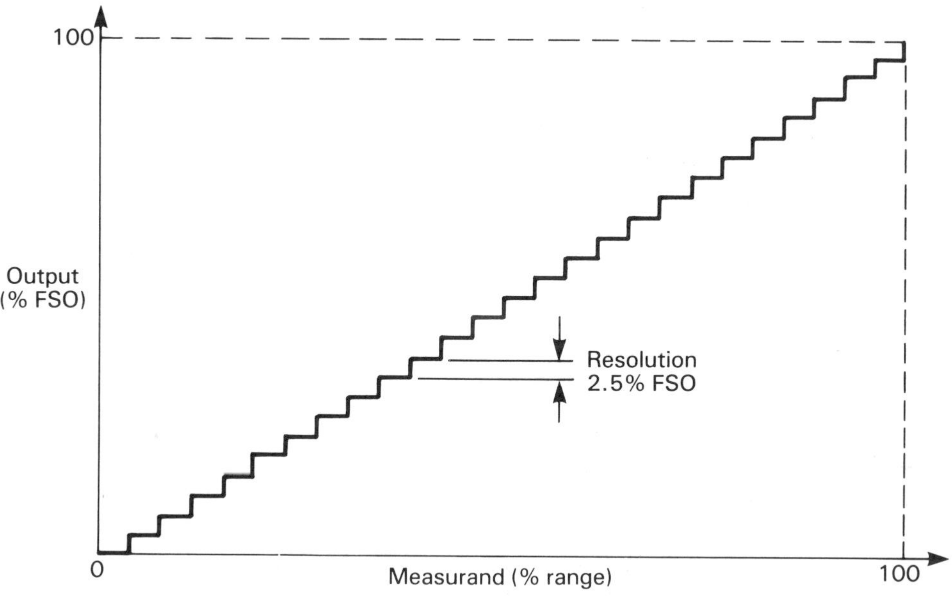

Figure 1.8

Dynamic characteristics relate the response of the device to variations of the measurand with respect to time. The most important ones are the frequency- and step-response characteristics.

Frequency response—can be defined as the variations in the output curve given a sinusoidally varying, constant amplitude, input measurand, see Figure 1.9. Please note that the output is normally shown as a ratio of the output amplitude divided by measurand amplitude, expressed in percentage.

Step response—can be defined as the variation in the output value given a

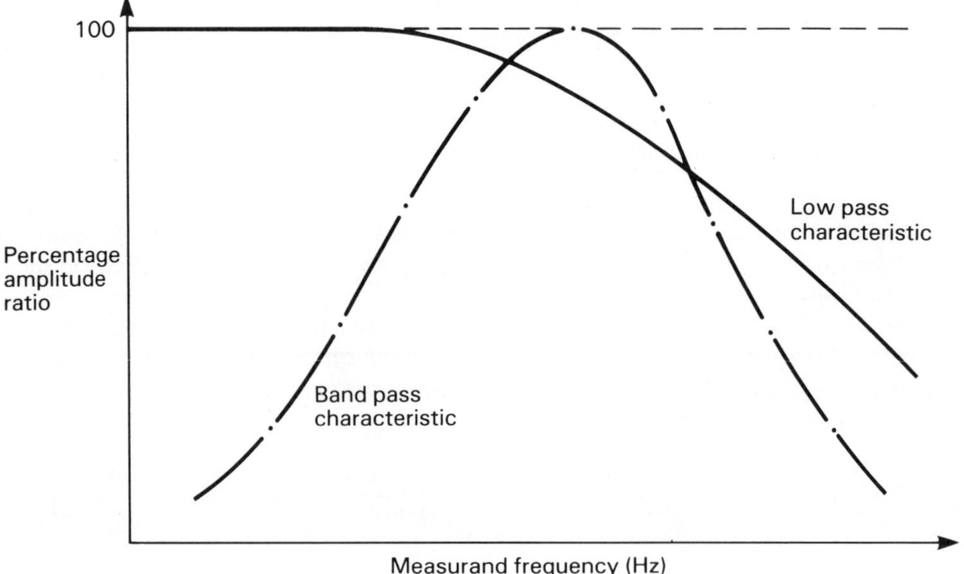

Figure 1.9

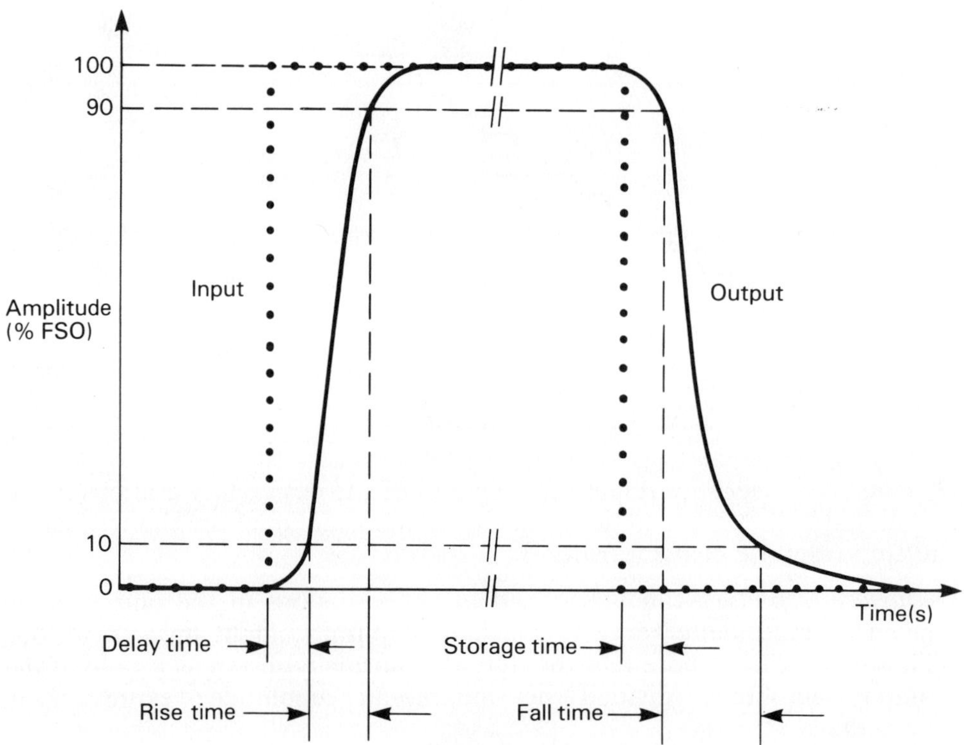

Figure 1.10

step change in the input measurand value. The time required for the output to reach a specified percentage of the final value is termed the 'response time'. These specified percentages are usually 10% and 90% of the final value.

As shown in Figure 1.10 there are four main types of response time:

- *Delay time*—the time taken for the output to rise to 10% of the final value once the input step has been applied.
- *Rise time*—the time taken for the output to rise from 10% to 90% of the final value.
- *Storage time*—the time taken for the output to fall to 90% of the final value once the input step has been removed.
- *Fall time*—the time taken for the output to fall from 90% to 10% of the final value.

Other important elements of the output characteristic are noise and noise margins, which can be defined as:

Noise—the level of any spurious signal(s) appearing at the output of the device due to any cause other than the driving input physical quantity.

Noise margin—the maximum noise level that can be tolerated by the device before it has any significant effect on the output signal.

PART I

TRANSDUCERS

Chapter 2

Position transducers

2.1 Overview

The control structure of a robot needs to know the position of each joint in order to calculate the position of the end effector (e.g. gripper) thus enabling the successful completion of the programmed task. The movements of the joints can be angular and/or linear depending on the type of robot; they are illustrated in Figure 2.1 for each of the four main robot coordinate systems. A suitable algorithm for calculating the end effector position in any of the coordinate systems can then easily be deduced, as shown in Figure 2.2 for the polar coordinates type robot, thus permitting the development of the necessary robot control software.

It should be noted that future developments in the robotics field may one day allow the use of external sensory feedback, or exteroceptors, as

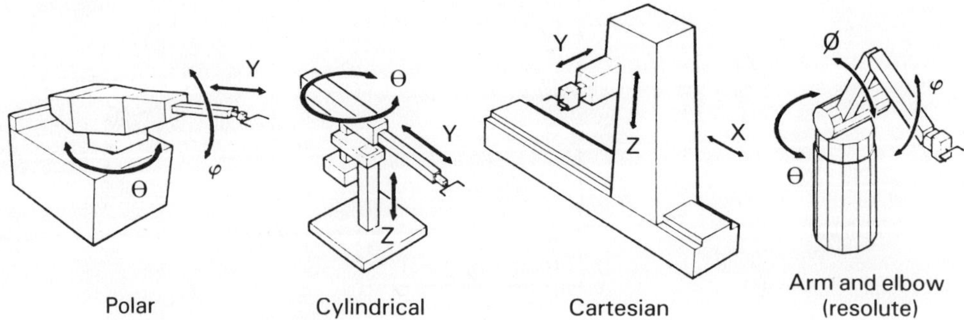

Figure 2.1 Robot coordinates systems (courtesy of PERA)

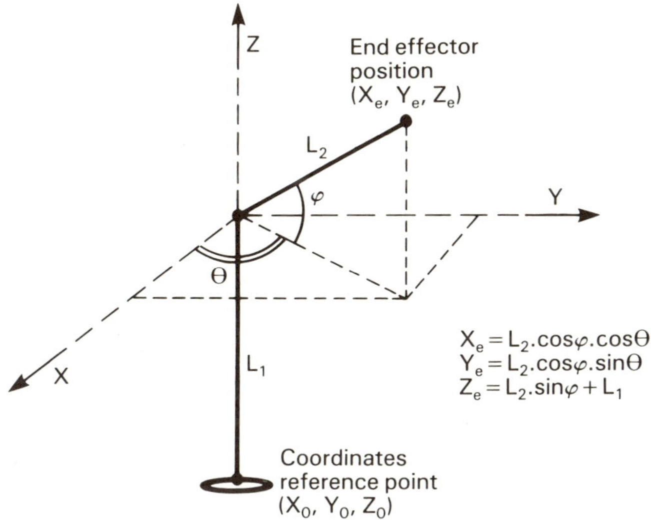

Figure 2.2 Schematic representation of polar coordinates robot and relative end effector position formulae

shown in Figure 2.3, to actually 'measure' the end effector position in relation to its surroundings, thereby dispensing with the need to calculate it from the joint positions. Indeed some commercial systems have successfully proved the worth of this technique in industrial applications. This approach, however, based on an adaptive technique similar to the way the human body

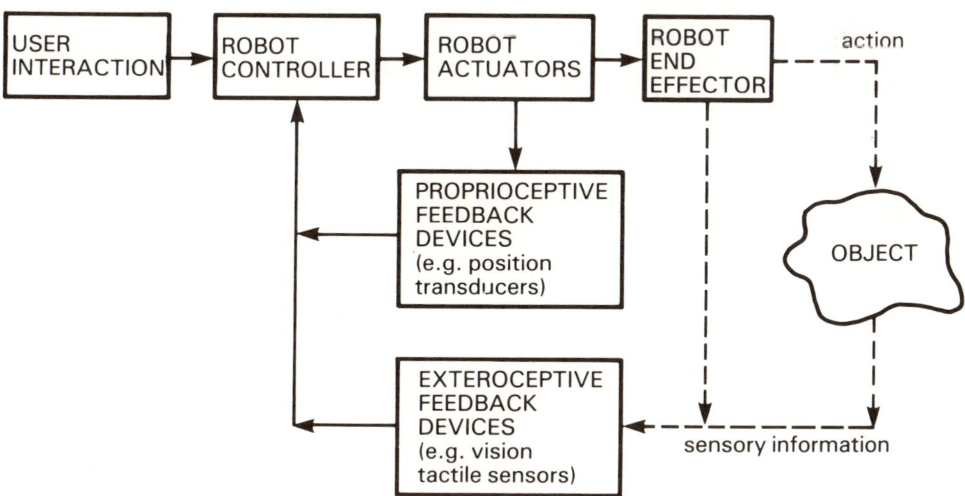

Figure 2.3 Block diagram of second generation robot system, showing the use of exteroceptive feedback

operates, is still largely at the research stages and may not be implemented generally on commercial machines for some time.

Internal position transducers (proprioceptors) therefore remain, at least for the present, the most accurate and reliable way of determining the end effector position within a robot control structure. There are two main types of position transducers: absolute and incremental (or 'relative'). The absolute position transducers are themselves further subdivided into resistive and optical types, as illustrated in Table 2.1. Other types of position transducer are also employed in industry, such as differential transformers and synchros, but their larger size and/or higher cost have precluded their use within robot structures and will not therefore be discussed here.

Table 2.1 Typical robot position transducers

Type	*Class*	*Device description*
Absolute	Resistive	Potentiometer
	Optical	Coded optical encoder disks
Incremental	Optical	Slotted optical encoder disks

2.2 Potentiometers

A potentiometer converts a mechanical input, namely the physical position of its wiper terminal, into an electrical signal by the simple principle of a potential divider, as illustrated by Figures 2.4 and 2.5 and eqns (2.1) and (2.2).

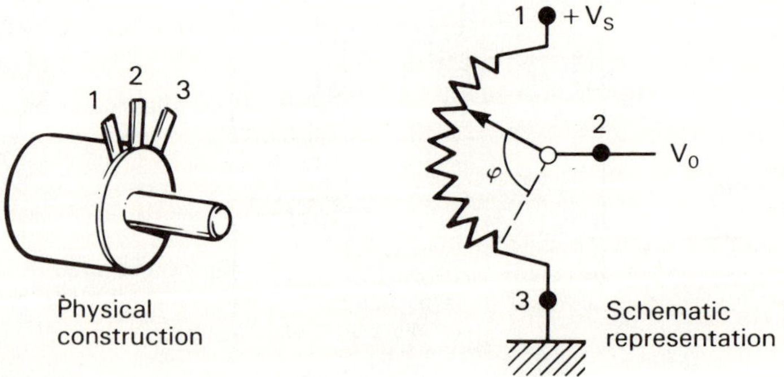

Figure 2.4 Angular position measurement using a potentiometer

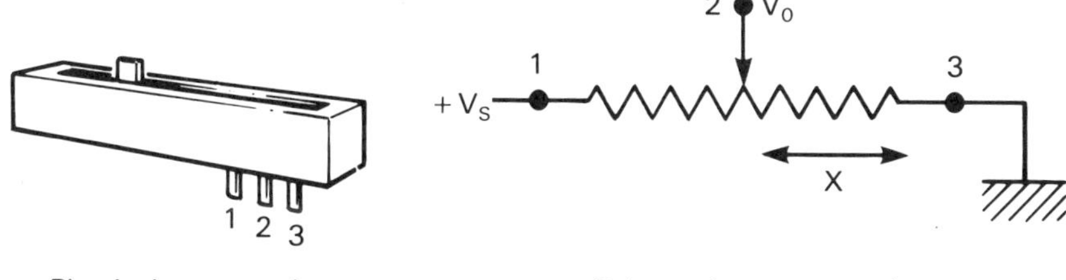

Figure 2.5 Linear position measurement using a potentiometer

$$V_0 = \frac{V_s R_\varphi}{R_{tot.}} = \frac{V_s \varphi}{\theta_{tot.}} = K\varphi \tag{2.1}$$

$$V_0 = \frac{V_s R_x}{R_{tot.}} = \frac{V_s X}{X_{tot.}} = KX \tag{2.2}$$

(Given that the supply voltage V_s, the total resistance value $R_{tot.}$, the total angular displacement $\theta_{tot.}$ and the total linear displacement $X_{tot.}$ are known constants.)

The resistive track can be made of a carbon film to reduce cost, or a cermet (conductive ceramic) film to increase the resolution and reduce noise, or a wound metal wire to allow higher power dissipation. The main advantages of potentiometers as position transducers are low cost, small size and versatility of operation (for example, it can easily provide logarithmic or quadratic functions), whereas the two main drawbacks are, firstly, that it is an inherently analogue device and therefore requires additional hardware, usually an analogue-to-digital converter (ADC), to interface it to a computer and, secondly, that its principle of operation requires the wiper to actually touch the main resistive medium, which makes it prone to mechanical wear and tends to limit its operational life.

Optical transducers are therefore slowly replacing the potentiometer in those position measuring applications which require very long operational life. Nevertheless the potentiometer is still widely used as a position transducer in most engineering applications in view of its ease of operation and versatility, small size and low cost.

An example of the potentiometer's popularity can be found in its use within the control structure of some educational and light industrial robots. The potentiometer's low cost and small size, in fact, allows full exploitation of the high torque–weight ratio of permanent-magnet d.c. motors, which has led some manufacturers to produce integral, lightweight and powerful pulse position (PWP) controlled d.c. servo units. For further details on such a device see Section 2.4 on Interfacing.

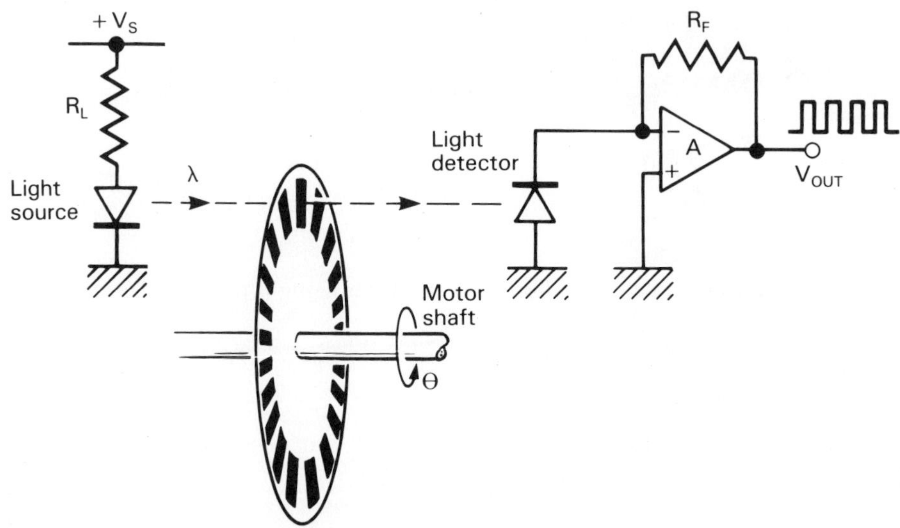

Figure 2.6 Generalized diagram of a position transducer system based on an optical encoder disk (incremental type shown)

2.3 Optical encoders

Optical encoders can be divided in to two main groups, as shown in Table 2.1, that is absolute and incremental transducers. Both types transform the mechanical input quantity, namely the physical angular position of its shaft, in to an electrical output quantity by light absorption. This principle is illustrated in Figure 2.6. The disk can be either transparent, such as clear plastic or glass, with opaque lines printed on it or made of metal with slots cut in it.

Optical encoders are widely used for the measurement of revolute joint positions in robot systems. Their main advantages are high accuracy (a 12 bit optical encoder will have an inherent reading accuracy of $1/4096 = 0.024\%$) and virtual absence of mechanical wear in operation due to the non-contact nature of the transducer. It should be noted that for each optical encoder there is a roughly equivalent magnetic position transducer, which also does not suffer from mechanical wear, but that the use of optical position transducers is much more widespread within robot systems.

However, this technique is somewhat inflexible because it requires prior knowledge of the total number of lines on the disk, so that the counting hardware can keep track of the number of complete shaft revolutions. Therefore changing the position measurement resolution, that is changing the total number of lines on the disk, would also require a change in the counting hardware.

An alternative, as adopted by most major manufacturers, is to use a third LED–photodiode pair and a separate channel containing a single

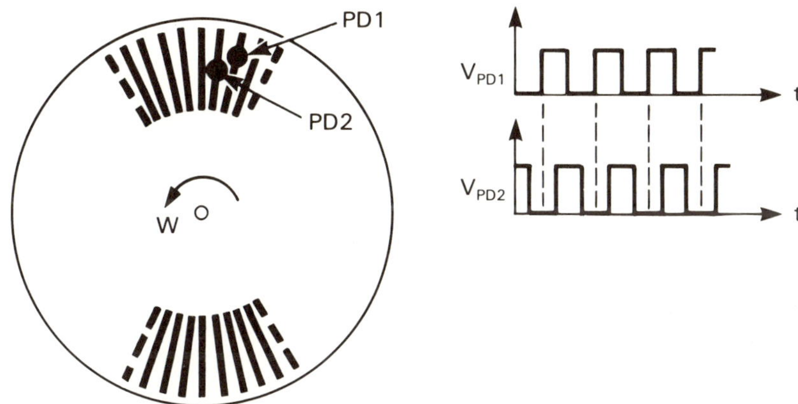

Figure 2.7 Typical position of angular direction sensing photodiodes

opaque line (or clear slot if a metal disk is used) which then provides the detection of each complete revolution, as required. The output signal from this third channel is in fact a pulse train whose period is equivalent to one complete shaft revolution; each pulse can be used to 'zero' the angular position count as well as providing the signal for a shaft revolution count, as shown in Figure 2.16 in the section on interfacing.

One drawback of the incremental encoder systems thus far described is their poor angular resolution, $\Delta\alpha$. This depends on the number n, of opaque lines on the glass disk (or the number of transparent slots on a metal disk), as shown in eqn (2.3), and the width W_p of the photodiode active area W_p on the disk plane, which in turn depends on the photodiode mounting distance from the disk centre, as illustrated in eqn (2.4) and Figure 2.8.

Note that, since most photodiodes have a square active area, the required dimension W_p is simply the square root of the active area given in

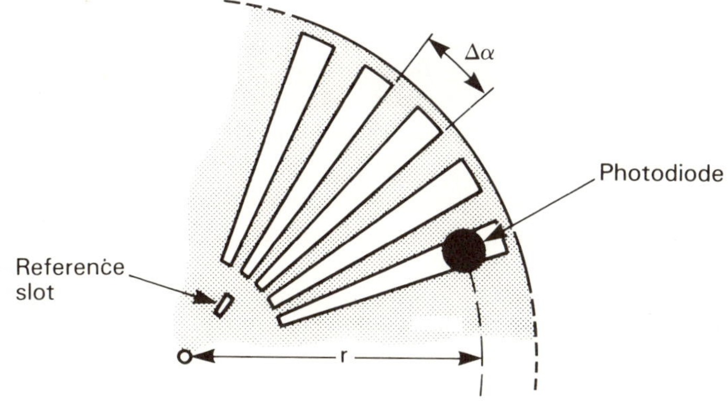

Figure 2.8 Angular resolution of optical incremental encoder

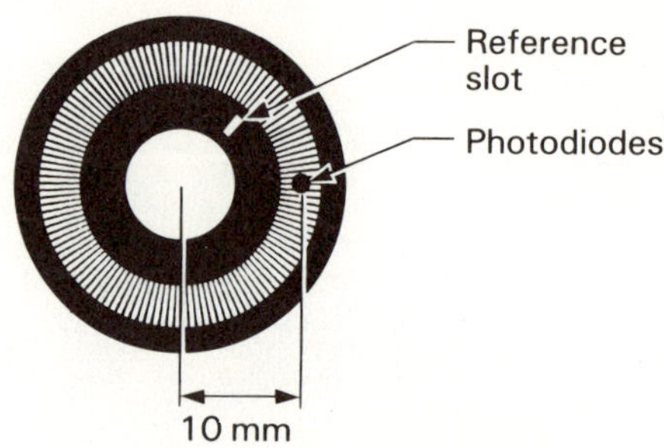

Figure 2.9 Incremental position transducer with 3° resolution

the photodiode data sheet:

$$\Delta\alpha = \frac{360}{n} \tag{2.3}$$

$$W_p = r \sin\left(\frac{\Delta\alpha}{2}\right) = r \sin\left(\frac{360}{n}\right) \tag{2.4}$$

For example: to design an optical incremental position transducer with a resolution of ±3° one would require an encoder disk with at least 120 slots, that is one every 3° of arc. To match this resolution the photodiode would need to havc an active width W_p no bigger than 0.25 mm² mounted 10 mm from the disk centre. An example of such a system is shown in Figure 2.9.

There are two main ways to increase the angular resolution without resulting in a larger disk diameter.

The first alternative relies on using gearboxes to yield *n* disk revolutions (where *n* is an integer >1) for every load shaft revolution. Since motors are often used in conjunction with gearboxes this alternative in inexpensive and only requires mounting the encoder disk on the same shaft as the motor instead of the load one. The encoder disk would therefore be rotating at the motor speed and the counting hardware would need to be designed to cope with the higher operating frequency; this is not usually a problem since a typical motor speed of 6000 rev/min would produce a pulse train of only 25.6 kHz at the output of a 256-line optical incremental position transducer, well within present digital technology operating frequencies. Any gear inaccuracies, such as backlash, would however, add to the load position measurement error, thus seriously limiting the overall transducer performance.

Another way to increase the optical resolution of these transducers is to use a second stationary 'phase plate' between the rotating disk and the photodiode so as to form Moiré fringes on the photodiode surfaces. Figure 2.10 shows the schematic diagram of a commercially available optical

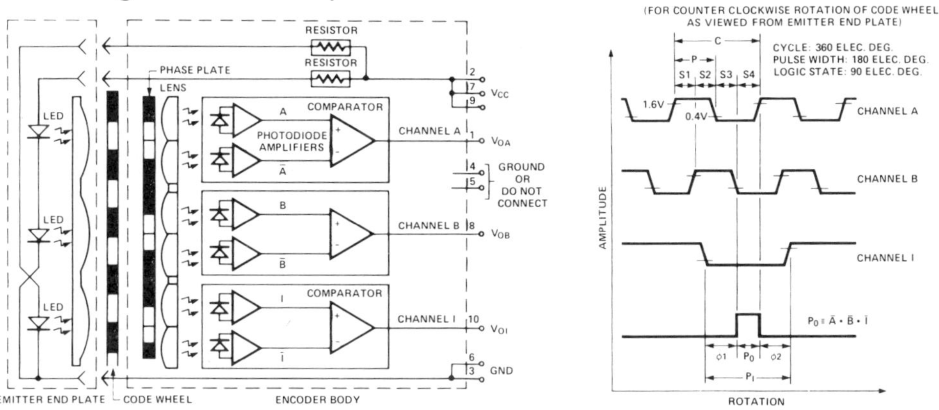

Figure 2.10 Hewlett–Packard incremental optical encoder (courtesy of Hewlett–Packard)

incremental transducer based on such a principle. The angular resolution of this device does not depend on the photodiode active area and its related optics but on the Moiré line width which can be adjusted to suit the application. Figure 2.11 shows a typical Moiré fringe pattern obtained using another commercially available device based on this principle.

2.3.1 Absolute encoders

Absolute position transducers are necessary when the control system is subject to frequent power shut-down periods and when frequent resetting of the robot arm positions needs to be avoided (as would be required if incremental types were used), such as in high flexibility and high reliability position sensing systems. Absolute optical encoders are used, for example, in the UNIMATE 2000 and 4000 series hydraulic robots.

Optical absolute encoder disks are based on a principle similar to that for incremental ones, producing an electrical signal proportional to the shaft angular position by light absorption using opaque lines on a transparent disk. The difference lies in the opaque pattern used—where the incremental disk uses a channel of alternate opaque and transparent lines, the absolute encoder disk has several channels of either binary or Gray coded patterns with each channel requiring an LED–photodiode combination to detect the logic state of each bit, as shown in Figure 2.12 for a 4-bit Gray coded disk.

The output therefore is provided directly as a digital number representing the angular position measurement whose resolution depends on the number of bits, that is the number of channels used, as shown by eqn (2.5),

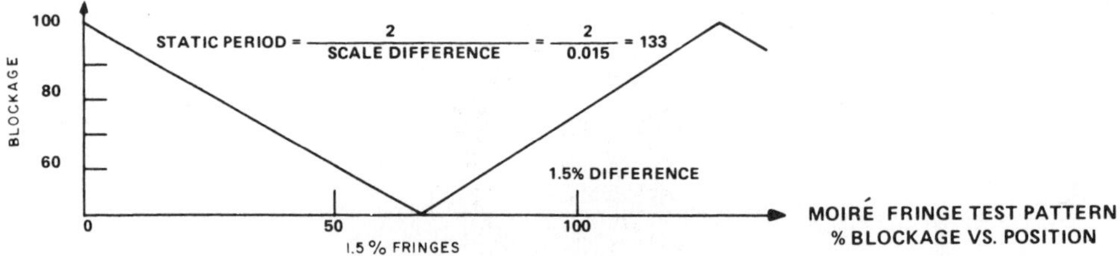

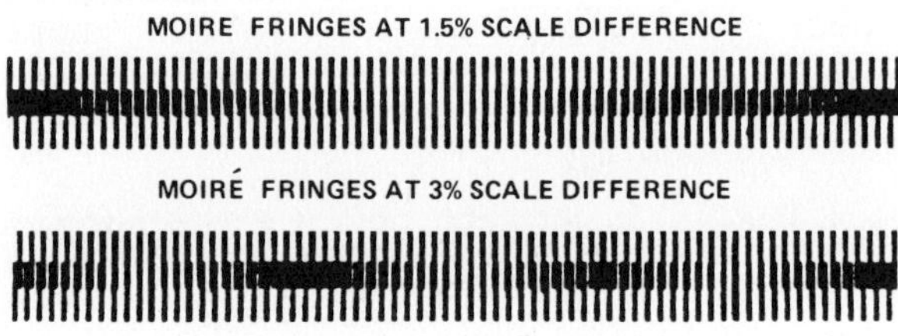

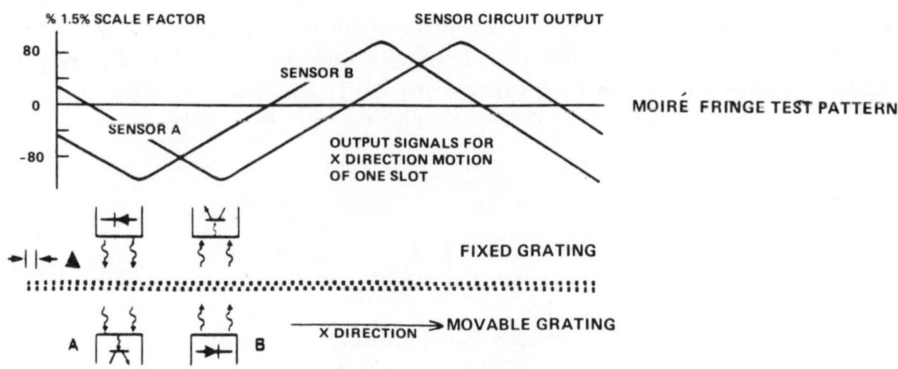

Figure 2.11 Optical position transducer based on Moiré fringe pattern (courtesy of General Electric)

where c is the number of channels required:

$$\text{Required angular resolution} = \frac{360}{2^c} \qquad (2.5)$$

For instance, to design an absolute optical position transducer with a resolution of less than 5° of arc one would need an encoder disk with seven channels, since $360°/2^7 = 360°/128 = 2.81°$ which is physically bigger and more expensive than an incremental encoder of comparable resolution.

The encoded disk pattern usually employed is based on a Gray code

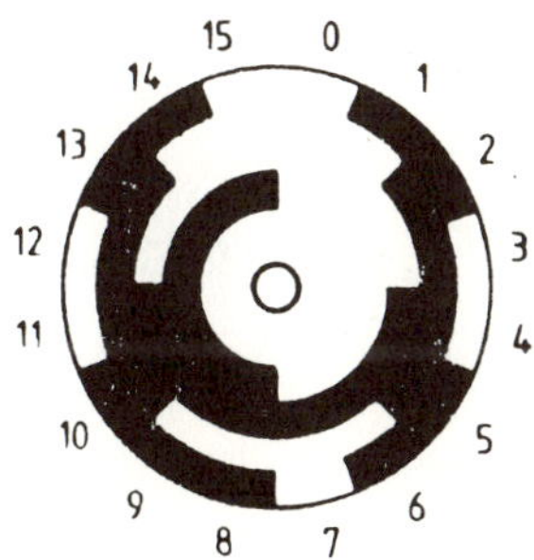

Figure 2.12 Absolute optical encoder based on a 4-bit Gray code (after McCloy and Harris, courtesy of Open University Press)

because it minimizes measurement errors since it allows only one bit to change state at any point in time. The alternative, namely a binary coded disk pattern, is not very popular because it may produce 'race' conditions in the digital counting hardware which would lead to measurement errors; this problem can occur in digital systems whenever more than one logic state changes at the same time, such as when counting from 01111111 to 10000000 when all 8 bits change. Figure 2.13 allows the comparison between a 4-bit absolute encoder disk based on Gray code and one based on binary code.

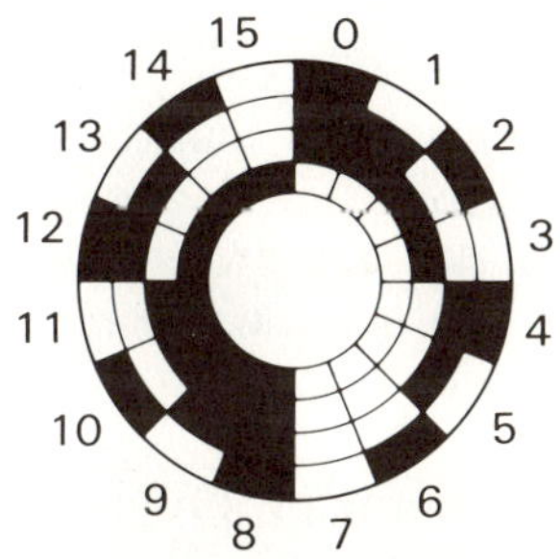

Figure 2.13 Absolute optical encoder based on a 4-bit binary code

2.4 Interfacing of position transducers

To interface an absolute position transducer to a computer is straightforward. The *coded optical encoder,* in fact provides a direct digital reading of the robot arm's absolute position relative to its end stops. The computer only needs to convert the transducer output code into binary code to use in all the subsequent robot control calculations. When a binary code disk is used, such a code conversion is, of course, unnecessary; in both cases,

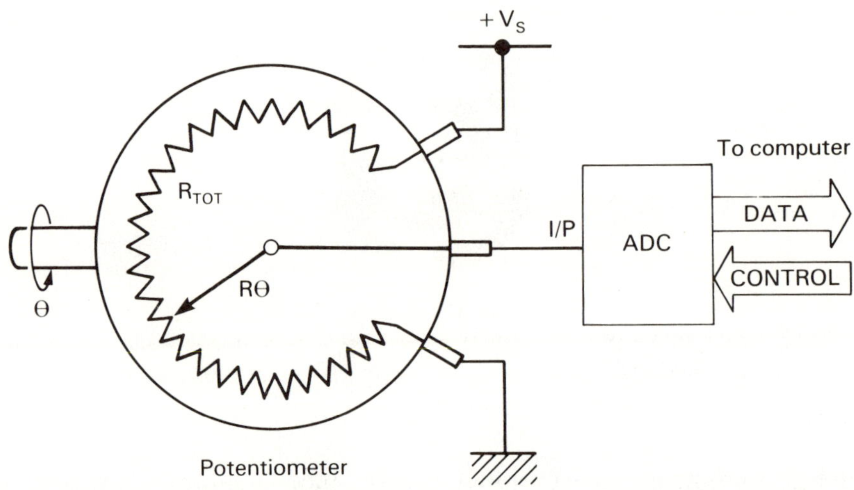

Figure 2.14 Schematic diagram of potentrometer interface circuit

however, there is no real need for hardware interfacing between the optical absolute transducer and the computer since any code conversion is usually done in software using a look-up table stored in ROM.

To interface the other absolute position transducer, namely *the potentiometer,* is also straightforward. As shown previously in Section 2.2, the output of this transducer is a voltage which represents the robot arm absolute position, which means it can be interfaced to the computer with a simple ADC, as shown in Figure 2.14.

An alternative to this technique can be found in small robots such as those encountered in the educational market and other distributed control systems such as remotely controlled model aeroplanes. The potentiometer in this case is integrated with the actuator, a small permanent-magnet d.c. motor, and some interface electronics to produce an integral, lightweight d.c. servo unit. The principle of operation of such a device is illustrated in Figure 2.15.

In these units the potentiometer provides the resistance value for a monostable *RC* charging constant; in other words, the monostable pulse output depends on the potentiometer shaft position. The monostable output is compared with the input signal provided by the microcomputer controller and an error signal is generated. The width of this error signal determines the length of the motor shaft rotation, that is the size of the adjustment necessary to maintain the shaft position under computer control.

This interface technique allows the computer to 'delegate' the control of the robot arm to these servo units therefore freeing it for other tasks, and is eminently suitable for use in robot systems with limited computer power such as microcomputer-controlled robots. However, the computer does not have a direct measurement of the robot arm position so that, in robot control terms, this is an open-loop system which is undesirable when

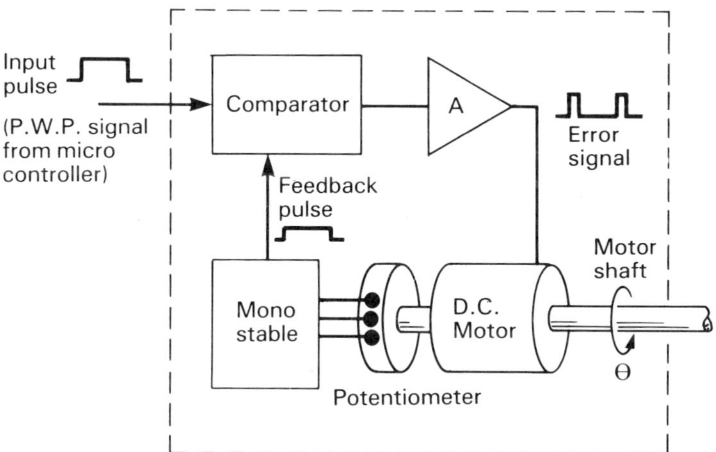

Figure 2.15 Pulse width position (PWP) controlled D.C. servo unit

requiring high positional accuracy such as in the case of robots handling complex tasks and/or expensive parts.

To interface an *incremental position transducer,* on the other hand, is more complex. The optical incremental encoder output is a low-voltage pulse train whose frequency is proportional to the shaft angular velocity. In order to measure the shaft position the interface hardware needs to count the number of pulses produced by moving from the last position. However, this latter position must be known at all times for the robot controller to be aware of all the joint positions within the working envelope, because this transducer can provide only an incremental (or relative) position measurement, that is it can show only how much a robot joint position has changed during the last move operation.

To overcome this problem the robot needs to be reset to a known position after switch-on (usually referred to as the 'home' position) and keep track of the joint absolute positions in the computer memory by updating it with the incremental position measurements after each move. Should the memory device storing these joint positions get corrupted for any reason, such as during a power supply surge, the robot would need to be reset again. The main components of the interface for this transducer are therefore an amplifier and a digital counter, as shown in Figure 2.16.

For reliable operation, however, this simple circuit requires the computer to estimate when the move operation has been completed (i.e. when the counter output has stopped changing) and to know which way the arm is moving (so that it can add or substract the last incremental position measurement from the previous one stored in memory), both of which are difficult to meet in practice because of possible arm oscillations caused by moving inertial loads.

In addition to the amplifier and the digital counter, a more practical

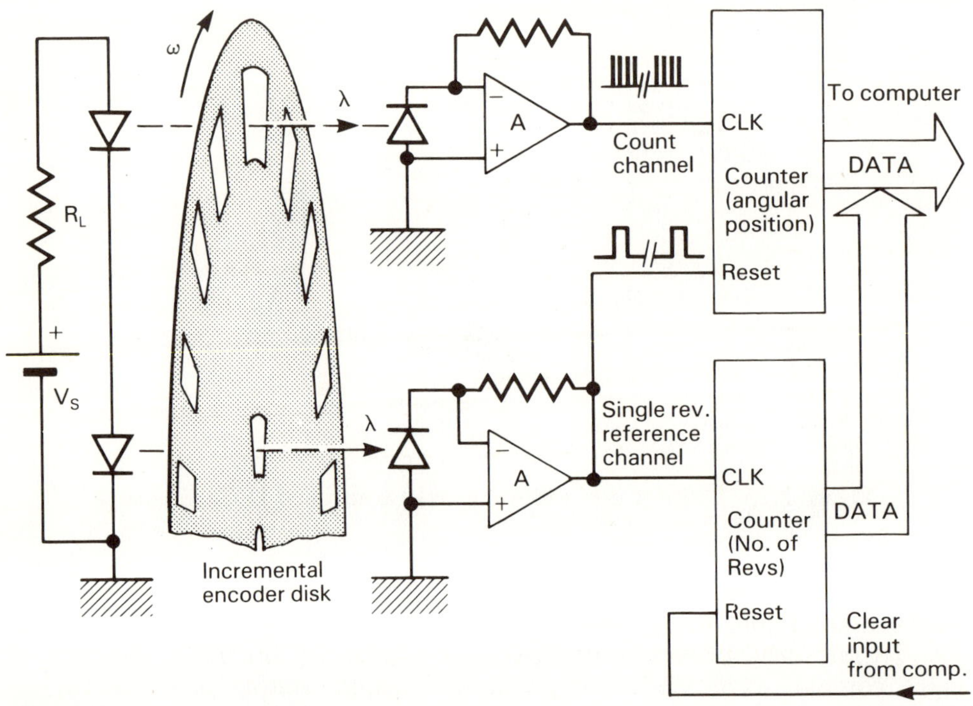

Figure 2.16 Example of computer interface circuit for optical incremental transducer system

interface circuit requires a form of local memory (so that the transducer output data can be held until the computer needs it) and a sequence detection circuit (to determine the direction of rotation by comparison of the two transducer outputs). An example of such a circuit is shown in Figure 2.17.

This is based on the high-resolution Moiré fringe type of optical encoder disk which provides directional movement information by virtue of the phase relationship of V_1 and V_2 as discussed previously in Section 2.3.1; note that, for clarity's sake, the 'number of revolutions' counter and associated components have been omitted from Figure 2.17.

The measurement resolution of the position transducer thus described depends on the line width of the Moire fringe pattern which can be obtained from the manufacturer data sheet; it should be noted, however, that the counter hardware must match this high resolution in terms of position count and may therefore need to be bigger than the 'standard' 8-bit width.

2.5 Conclusions

The optical incremental encoder, whether based on the Moiré principle or on a simple pattern of alternate opaque and transparent lines, is rapidly becoming

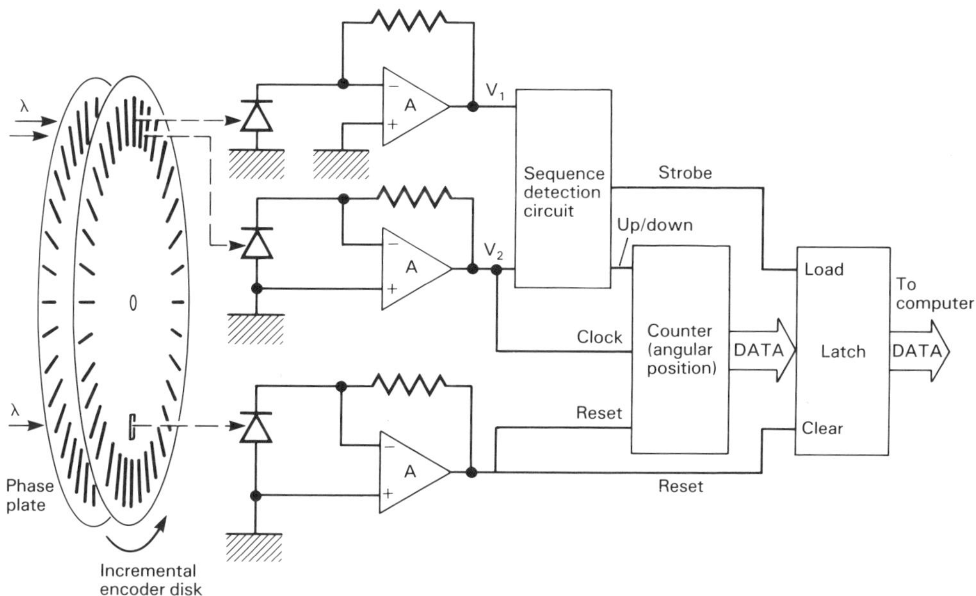

Figure 2.17 Example of computer interface circuit based on an incremental encoder capable of direction detection high resolution

the most popular position transducer in the field of machine control and robotics, replacing the potentiometer in many applications requiring longer operational life.

Optical absolute position transducers have been restricted in their use to high reliability, non-space-critical applications in view of their larger size and cost.

2.6 Revision questions

(a) Compare a wirewound potentiometer with a 120-line optical incremental transducer in terms of angular resolution, operational life, cost and size.

(b) Design an incremental position transducer system with a resolution better than 6° of arc and draw a suitable interface circuit diagram of your system.

(c) Draw an absolute encoder disk pattern to give an angular resolution less than 15° using binary coding.

2.7 Further reading material

Bannister, B. R. and Whitehead, D. G. (1986) *Transducers and Interfacing,* Wokingham, Van Nostrand Reinhold (UK).

Polet, T. W. (1985) *Industrial Robots: Design, Operation and Application,* Glentop Publishers.

Seippel, R. G. (1983) *Transducers, Sensors and Detectors,* Reston Publishing Co.

Snyder, W. E. (1985) *Industrial Robots: Computer Interfacing and Control,* Englewood Cliffs, Prentice-Hall.

Chapter 3

Light transducers

3.1 Overview

Light transducers convert electromagnetic energy inputs in to electrical energy outputs. The wavelength range of this input energy, as shown in Table 3.1, can be assumed to be 100 nm to 1000 μm.

For the purpose of this book, however, we shall restrict our interest to the visible spectrum as well as the nearby sidebands both in the IR and UV regions which, as shown in Figure 3.1, represents a range of 300 nm to 1500 nm (0.3 μm to 1.5 μm).

Most robot light transducers in fact work in the visible spectrum (e.g. cameras) but the use of solid state silicon devices as optical transducers (e.g. photodiodes) also extends the useful optical range into the near IR region. Vacuum photoemissive transducers (e.g. photomultipliers) enable the use of the higher energy content of ultraviolet radiation.

3.1.1 Photoelectric effect

Light transducers perform the energy conversion by absorption; that is the material atoms absorb the photon energy and use it to move one or more suitable electrons to a higher energy level, either within the valence band (i.e. to a higher atomic shell) or, by crossing the energy gap, in to the conduction band (i.e. leave the atom thereby raising the material conductivity). This phenomenon, illustrated in Figure 3.2, is known as the Photoelectric Effect. There are 3 main types of photoelectric effects (Chappel, 1976):

Table 3.1 Subdivision of the optical radiation spectrum according to DIN 5031 (courtesy Texas Instruments)

Wavelength range	*Designation of radiation*
100 nm–280 nm	UV-C
280 nm–315 nm	UV-B
315 nm–380 nm	UV-A
380 nm–440 nm	Light-violet
440 nm–495 nm	Light-blue
495 nm–558 nm	Light-green
580 nm–640 nm	Light-yellow
640 nm–750 nm	Light-red
750 nm–1400 nm	IR-A
1.4 μm–3 μm	IR-B
3 μm–1000 μm	IR-C

(*i*) *Internal photoeffect*
This takes place when the conductivity of the bulk material is raised by the creation of electron-hole pairs (the shifting of electrons in to the conduction band in fact leaves holes back in the valence band). This is the basis for photoconductive transducers such as photoresistors.

(*ii*) *Junction photoeffect*
This is also known as the Photovoltaic Effect and takes place when the conductivity of the material is raised by the creation of electron-hole pairs within the depletion region (and within one diffusion length) of a P–N junction. In principle this is also an internal photoeffect but related to the physics of semiconductor P–N junctions and forms the basis for photovoltaic transducers such as photocells.

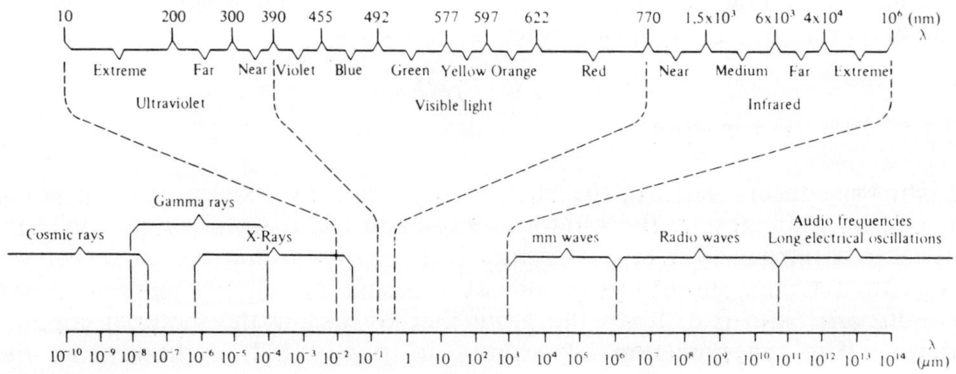

Figure 3.1 Electromagnetic radiation spectrum (courtesy Texas Instruments)

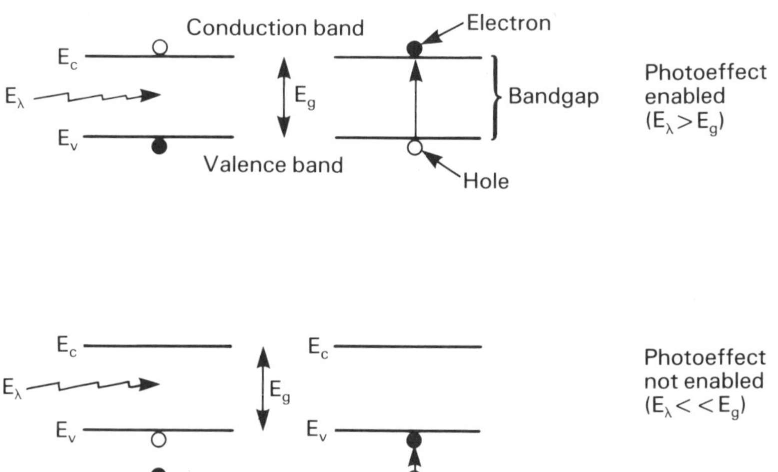

Figure 3.2 The photoelectric effect

(iii) External photoeffect
This occurs when the electron is given sufficient energy by the incident photon(s) to overcome the photoelectric threshold value and leave the material, thereby becoming a 'free' electron. This is the basis for photoemissive transducers such as photomultiplier and vacuum camera tubes.

It follows therefore that for any photoelectric effect to take place the incident photon(s) must have sufficient energy to raise the electron(s) to the required level. The energy content of a photon w is given by its wavelength λ, its speed of propagation—which for most purposes can be assumed to be approximately the same as the speed of light in vacuum—c and Plank's constant h, as shown in eqn (3.1):

$$\omega = \frac{h \cdot c}{\lambda} \tag{3.1}$$

3.2 Photoresistors

Photoresistors, which are sometimes also referred to as Light Dependent Resistors (LDR), are based on the internal photoeffect, are made mostly from semiconductor material such as Si, Ge and CdS (in both the intrinsic and doped form) and have a logarithmic transfer characteristic, as shown in Figure 3.3 for a typical CdS photoresistor.

Because of the relatively long life time of the optically induced charge carriers (i.e. the time taken for an electron in the conduction band to recombine with a hole in the valence band) and the dependence of this

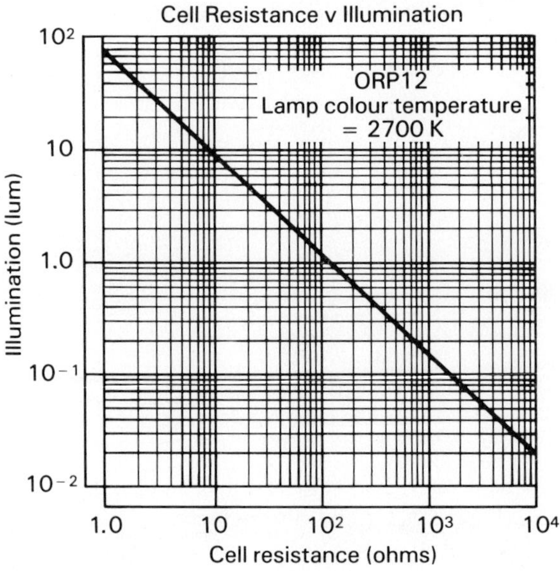

Figure 3.3 Photoresistor ORP12 transfer characteristic

process on previous conditions (i.e. a large hysterisis), photoresistors are used mainly in low-frequency control applications where their slow response time and relatively high hysteresis are either acceptable or an actual advantage.

Photoresistors therefore find limited applications in the robotics field where they are used mainly in components detecton on conveyor systems or as part of an optical safety 'curtain' around industrial robots.

3.3 Photodiodes and photocells

These are photovoltaic devices based on the junction photoeffect, that is the creation of optically induced electron-hole pairs within a P–N junction. It is therefore useful at this point to retrace some of the basic steps of semiconductor physics.

Electron-hole pairs are generated by thermal agitation throughout a semiconductor material and are, by their nature, in continuous random movement. Upon formation of a P–N junction (e.g. by diffusion of an element of opposite type to that of the bulk dopant) some of the mobile carriers from one side cross the junction and recombine with some of the mobile carriers on the other side (i.e. electrons from the N region recombine with holes from the P region) leaving behind fixed positive and negative ions. The process continues and this thin region either side of their junction, now depleted of mobile carriers (and thus termed 'depletion' region), widens until the electrical potential created by the fixed ions is sufficient to prevent any further diffusion (i.e. junction crossing) by the mobile carriers. The junction is now in thermal equilibrium and has acquired a so called 'barrier potential' (this however cannot be measured since it is cancelled out by the external connections contact potential). If the material is now irradiated, any electron-hole pairs produced by the internal photoelectric effect within the depletion region are swept across the junction by the electric field associated with the barrier potential thereby producing a current flow, as illustrated in Figure 3.4.

This injection of optically induced majority carriers therefore reduces the barrier potential and, since the connections contact potential is still the same, it produces a measurable voltage across the junction. The device thus created is therefore an electrical generator capable, as shown in Figure 3.5, of driving a current through a suitably connected external load, with the P terminal becoming positive with respect to the N terminal.

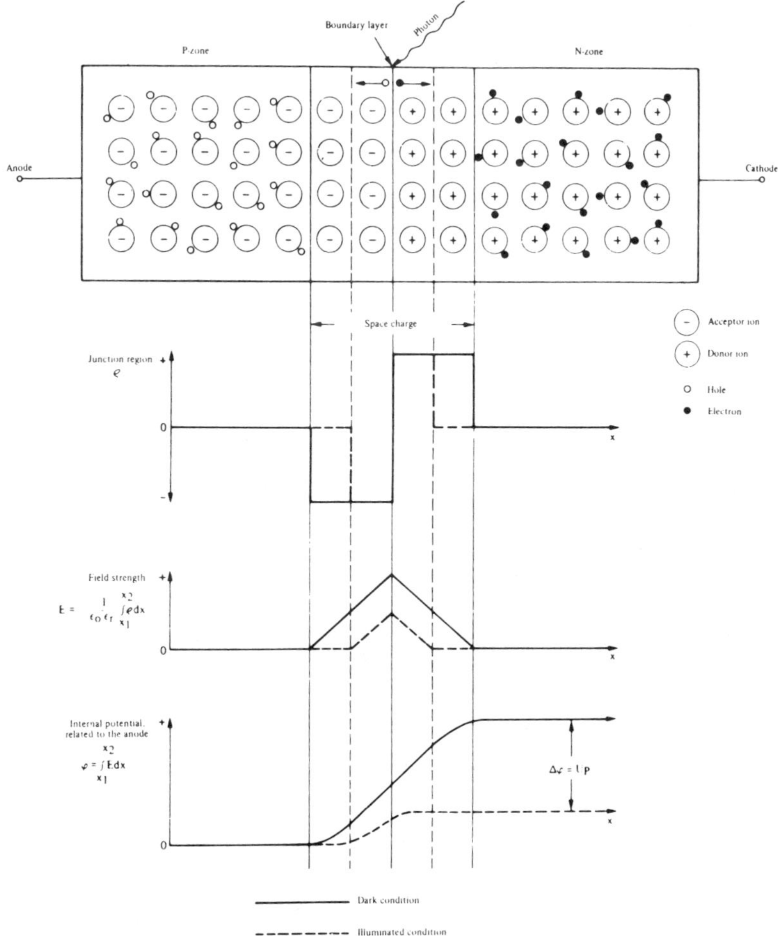

Figure 3.4 Photovoltaic transducer principle (courtesy of Texas Instruments)

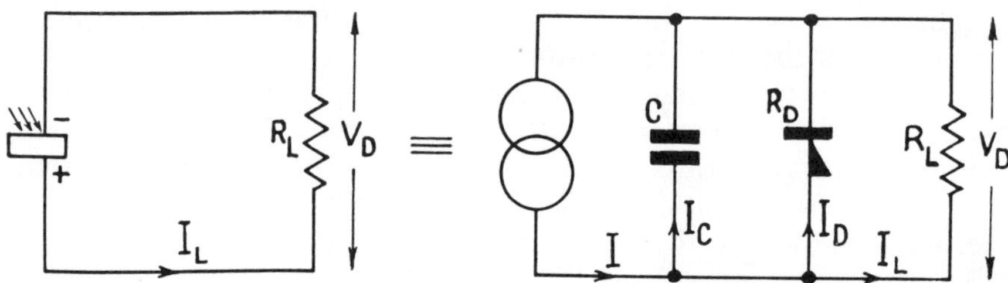

Figure 3.5 Photocell diagram and equivalent circuit (courtesy of Ferranti)

The photocurrent which has been caused to flow across the junction (i.e. the rate of change of the majority carriers crossing it) is approximately proportional to the rate at which the light quanta impinge on the photovoltaic device and therefore it increases with the intensity of the illumination. This is shown diagrammatically in Figure 3.6 under different lighting conditions.

When the device has no load connected to it the external photocurrent must, naturally, be zero and therefore all the current flows through the diode in the forward direction. This produces an 'open circuit voltage' V_{oc} which is a logarithmic function of the photocurrent I_p and has an upper limit of 0.6 volts for silicon devices (a photovoltaic device used with a very large load resistor will therefore act as a logarithmic voltage source). When the device is short circuited, on the other hand, the maximum current that can flow from the P to the N region is the short-circuit current I_{sc} which is equal to the photocurrent I_p (a photovoltaic device used with a small load resistor

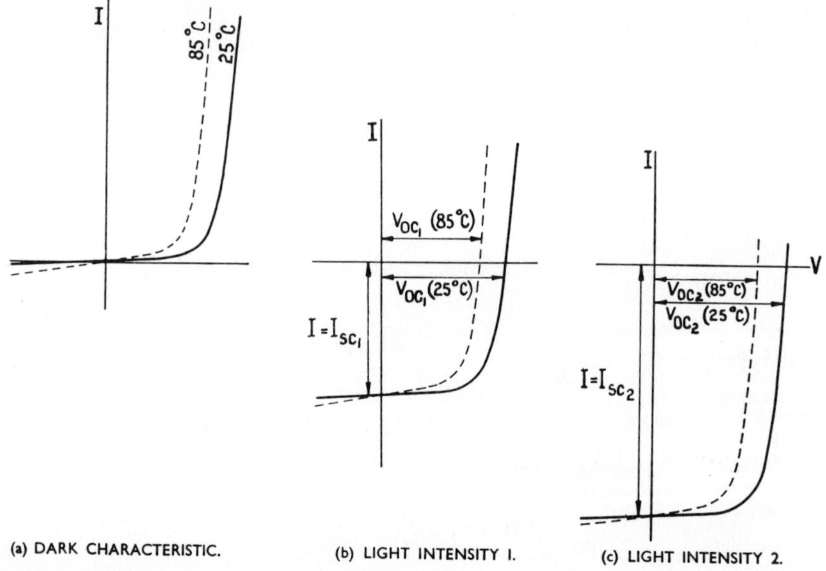

Figure 3.6 Photocell V/I curves under different illumination (courtesy of Ferranti)

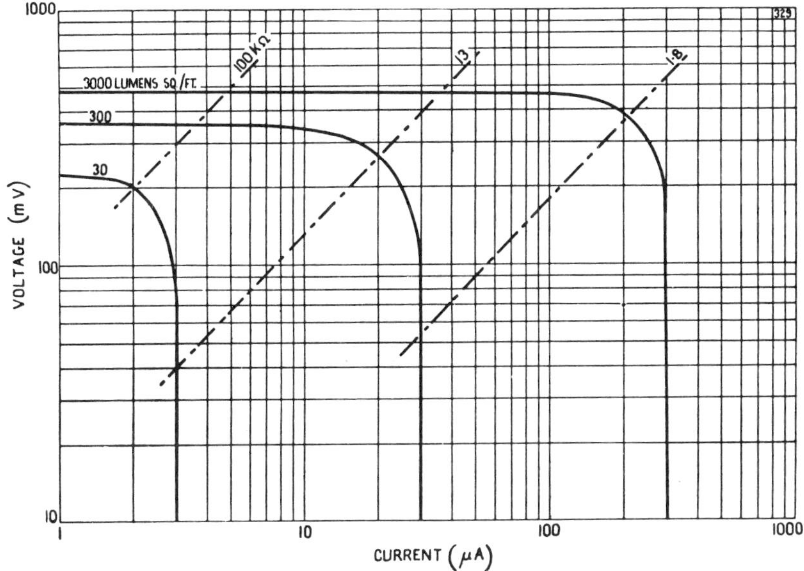

Figure 3.7 Typical photocell voltage/current curves (courtesy of Ferranti)

will therefore act as a linear current source). This behaviour is illustrated by re-drawing the fourth quadrant of the device I–V characteristic (shown in Figure 3.6) to highlight the open and short circuit operating points under different illumination conditions, as shown in Figure 3.7. This is the principle of operation of a silicon photocell.

The operation of a *photodiode* is very similar, indeed it differs only in that it is used with a reverse bias. This means that a larger depletion region is formed either side of the junction and only a small reverse current flows in the absencc of any illumination (the so called 'dark' current). When the junction is irradiated by light of a suitable wavelength, the additional carriers thus created raise the reverse current by a value proportional to the incident radiation therefore yielding similar curves to those of the photocell (as shown previously in Figure 3.6, third quadrant).

A photodiode can therefore be used as a photocell in the absence of an external bias but exhibits a smaller junction capacitance under reverse bias (i.e. a wider gap between the virtual 'plates' of the stray capacitor) and is therefore suitable for high-speed applications such as light pulse measurement, as required in optical rangefinder systems (see Chapter 6 for further details). The photodiode frequency response is in fact governed by its effective load resistor and the junction capacitance, as shown in eqn (3.2):

$$f_{bw} \simeq \frac{1}{2 \cdot R_{\text{load}} \cdot C_{\text{junction}}}. \tag{3.2}$$

As well as the basic junction photodetectors described so far, such as the photocell and photodiode, there are other more sensitive devices also

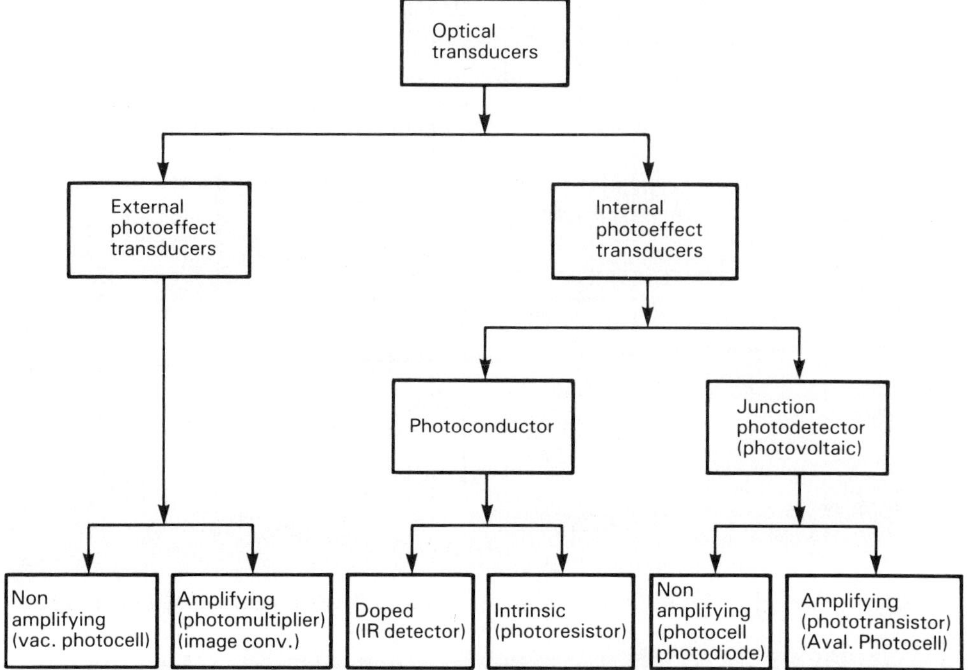

Figure 3.8 Optical transducers summary

based on the junction photoeffect but which provide a higher current output for the same illuminance input value. These devices, in other words, have an amplification function associated with the basic junction photoeffect and, as shown in Figure 3.8, are known as *amplifying photodetectors.* The most common examples of such devices are the phototransistor and the avalanchephotodiode.

3.3.1 Phototransistors

The phototransistor, as illustrated in Figure 3.9, is a photodiode-transistor combination, that is the base current of the transistor is generated by irradiating the reverse biased base-collector junction which acts as an in-built photodiode.

The output current (I_c or I_e, depending on the configuration) therefore relates to this base photocurrent I_p via the d.c. current amplification h_{FE}, as shown in eqn (3.3):

$$I_c = I_p * h_{FE} \quad \text{or} \quad I_e = I_p * (h_{FE} + 1). \tag{3.3}$$

For some applications the base may need to be biased to a specific operating point (e.g. to operate the transistor in a more linear part of the transfer characteristic) in the which case the transistor output current will be

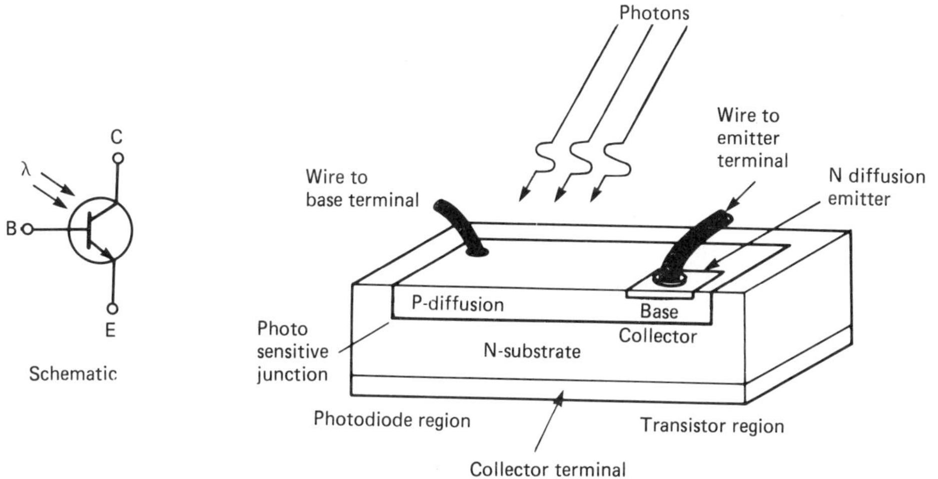

Figure 3.9 Phototransistor (courtesy General Electric)

proportional to the algebraic sum of the bias current I_B and the photocurrent I_p.

Phototransistors are therefore more sensitive than photodiodes but do have worse output current tolerances and a less linear transfer characteristic due to the h_{FE} variations with temperature, bias voltage and collector current. The speed of response of a phototransistor also differs from that of a photodiode due to the dominant effect of the input R–C time constant on the output rise and fall times (i.e. the bandwidth of a phototransistor is governed by the effective 'input' R–C value rather than the actual transistor switching speed) as well as the effect of the current amplification factor h_{FE}, as shown in eqn (3.4):

$$\text{Output time constant} = T = h_{FE} \cdot R_{IN} \cdot C_J. \quad (3.4)$$

Phototransistors are therefore mostly used in low cost, medium speed and medium sensitivity light detection but are slowly being replaced by photodiode-operational amplifier combinations due to the market availability of low cost, high gain operational amplifiers whose performance can be selected to suit the application.

3.3.2 Avalanche photodiodes

Avalanche photodiodes also provide higher output currents than ordinary photodiodes by using avalanche multiplication (in electronics 'avalanche' is in fact defined as cascade multiplications of ions) to amplify the photocurrent I_p created by the junction photoeffect, as shown in eqn (3.5):

$$I_d = I_p \cdot M \quad (3.5)$$

The value of the avalanche multiplication factor M, however, suffers from statistical variations (as does any avalanche multiplication device—see also Photomultipliers) and depends on the stability of the applied reverse voltage V_R at the avalanche voltage V_a (for $V_R < V_a \rightarrow M = 1$ whereas for $V_R > V_a \rightarrow M \gg 1$). To achieve a high signal-to-noise ratio, therefore, a point on the characteristic must be found where M is sufficiently large (i.e. V_R as large as possible) but also where the avalanche current is due mainly to light generated carriers and not field induced carriers (i.e. V_R as small as possible). The balance between noise and gain is thus difficult to achieve, temperature stability is also poor and a highly stable bias voltage source is required (typically it needs to be stable to within 0.1% of its set value) thereby raising the cost and limiting its applications. The avalanche photodiode is, nevertheless, a very sensitive high frequency device (due to the high speed characteristic of the avalanche effect) and is therefore suitable for wide dynamic range, broadband light amplification such as is the case in the front-end of optical range finders (Chappel, 1976).

3.4 Photomultipliers

The photomultiplier is a vacuum photoemissive device and is based on a photocathode, a series of electron multiplying electrodes called dynodes and an anode, as shown in Figure 3.10:

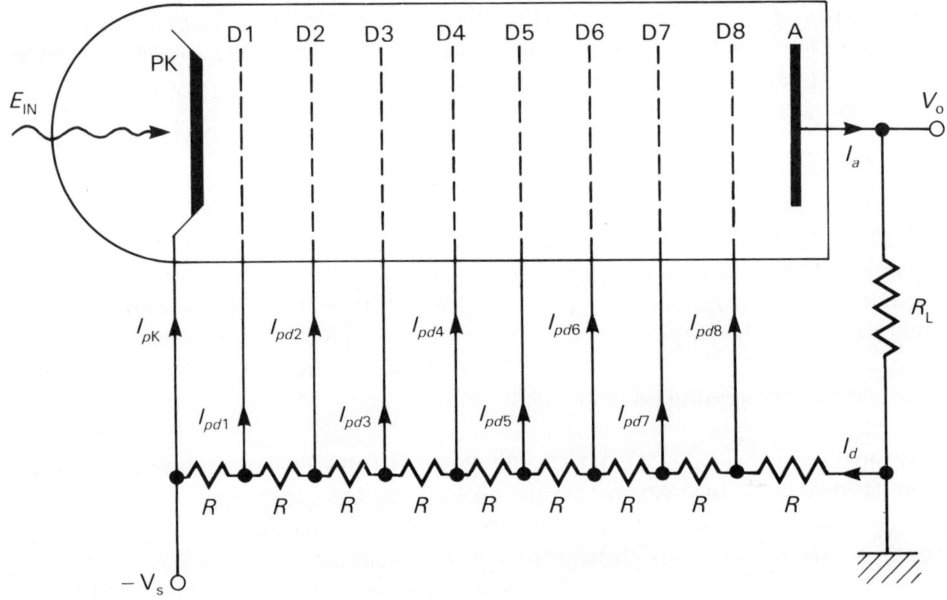

Figure 3.10 Photomultiplier

The photocathode is a photosensitive surface consisting of a support (usually transparent to the incident radiation) coated with an opaque material whose 'photoelectric threshold value' (i.e. the electron escape energy) is sufficiently low to allow a portion of the incident radiation energy to be used to 'free' electrons from the photocathode material (i.e. enable the external photoeffect to take place). The nature of the photocathode coating material therefore depends on the energy range of the expected incident radiation, that is on the incident light wavelength.

The free electrons thus produced by the photocathode are accelerated by the applied bias voltage towards the dynodes which are coated by a material with a low work function, so as to enable secondary emission and therefore current multiplication. The output current collected at the anode I_a is therefore related to the initial phototocathode current I_{pk} by a current amplification factor G, as shown in eqn (3.6):

$$I_a = G * I_{pk} \tag{3.6}$$

this latter depends on the number of dynodes n (i.e. the number of stages of amplification), the dynode bias voltage V_d and the dynode material/structure factor α (i.e. the secondary emission efficiency), as shown in eqn (3.7):

$$G = [A * (V_d)^{\alpha}]^n \tag{3.7}$$

where A is a constant and α is usually between 0.7 and 0.8 (Hamamatsu, 1983). Equations (3.6) and (3.7) show that the output anode current I_a is a linear function of the incident radiation E_{in} if the gain G is kept constant. This, according to eqn (3.7), can be achieved by keeping the dynode voltage V_d constant. The value of the supply voltage and the design of the dynode bias network is therefore critical both to the current gain actual value and its linearity. The power supply needs to be a stable, low ripple voltage source and the dynode bias needs to be insensitive to the voltage variations produced by the dynode photo-current I_{pd} (i.e. the dynode current needed to replace the electrons used in the photocurrent amplification). A simple potential divider network between the photocathode, the dynodes and the anode is sufficient but, for most applications, a voltage stabilizing zener diode and pulse decoupling capacitors also need to be used (Hamamatsu, 1983; Ferranti, 1982).

Like all vacuum components the photomultiplier is not suitable for applications where large mechanical forces are likely to be exerted on its envelope, furthermore its gain is affected by magnetic fields thus making the photomultiplier unsuitable for applications near (or in) strong magnetic fields without proper shielding (note that both drawbacks can be eliminated by the use of high permeability metal casings). However, in view of its wide dynamic range of operation (typically 4 to 5 decades) and its high speed characteristic (typical value for the anode current rise time is 2 nS, limited largely by the electrons' transit time and the electrodes stray capacitances) the photomultiplier is ideal for high sensitivity, high speed light amplification, such as in the case of Scanning Laser Rangefinders.

3.5 Optical array transducers

As well as the basic devices thus described, there are others that embody more than one elementary transducer in their construction and are thus referred to as 'composite' or, more specifically, 'array' transducers. These devices are in fact based on arrays of photosensitive elements known as photosites, with each array element working on the same principles as the elementary transducers described in the previous paragraphs. The array can be two dimensional, and is then known as an *area array* or area imaging device (AID) as shown in Figure 3.11, or it can be one dimensional in which case it is called a *linear array* or linear imaging device (LID).

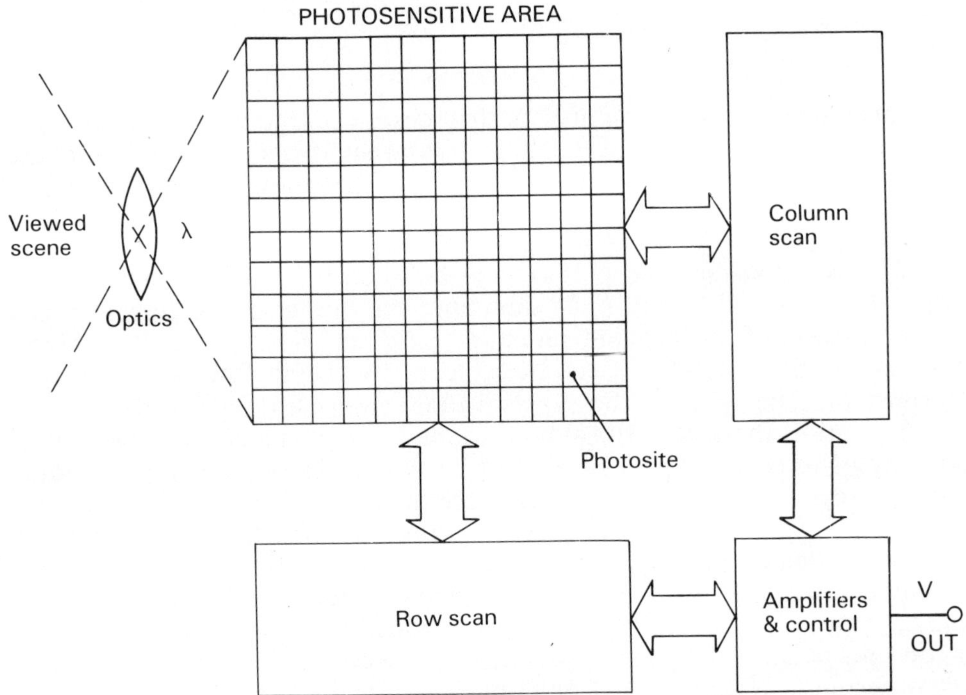

Figure 3.11 Optical array transducer

The technology used in the construction of these array transducers also allows a natural subdivision in to: vacuum and solid state devices which are more commonly known as, respectively, vacuum and solid state cameras, Solid state and, to a lesser extent, vacuum cameras are growing in importance in the automation and robotics fields due to their increasing use in applications requiring non-contact inspection and measurement (Batchelor, 1985; Pugh, 1983); this is particularly true, as will be shown in Section

3.5.2.1, for solid state linear array transducers (sometimes referred to as *linescan cameras*) which are extensively used to measure edge positions in such diverse applications as arc welding path control (Drews *et al.*, 1986), food manufacturing process control (Electronic Automation, 1985; Hollingum, 1984) and automated car production (Austin Rover 1984; Hollingum, 1984).

3.5.1 Vacuum optical array transducers

Vacuum optical array transducers are based on a 2-D photosite array, that is there is no vacuum 1-D equivalent of the solid state linescan camera, and have a typical construction as shown in Figure 3.12. These devices include

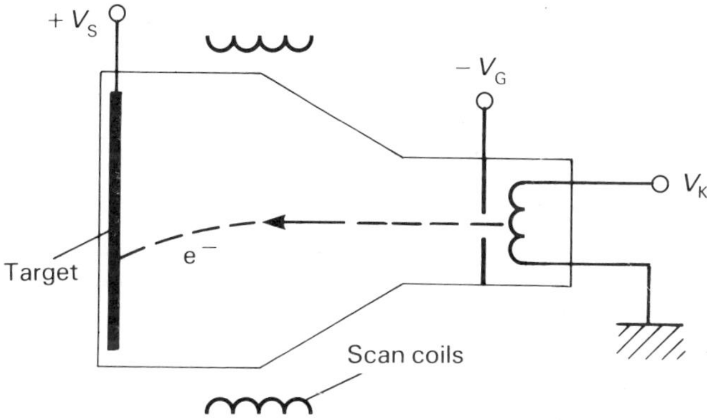

Figure 3.12 Schematic diagram of vacuum optical array transducer (television camera)

the television cameras and have traditionally been very popular as optical transducers even though, as shown later in Section 3.5.3, they are increasingly being challenged by solid state cameras in the robotics field. Vacuum television cameras include the recently introduced high sensitivity Newvicon and Chalnicon as well as the longer established Vidicon and Plumbicon cameras. These latter two types are by far the most popular and will therefore be described in more detail in this Section.

The main difference between the various vacuum television cameras lies in the photosensitive area (usually known as the 'target') which is coated with lead oxide (PbO) for the Plumbicon, selenium or antimony compounds (e.g. Sb_2O_3) for the Vidicon and zinc and cadmium tellurides for the Newvicon and Chalnicon.

The scanning of the photosites is achieved via an electron beam in the same way as in cathode ray tubes, namely by the use of magnetic deflection coils. The principle of operation, illustrated by Figure 3.13, is as follows:

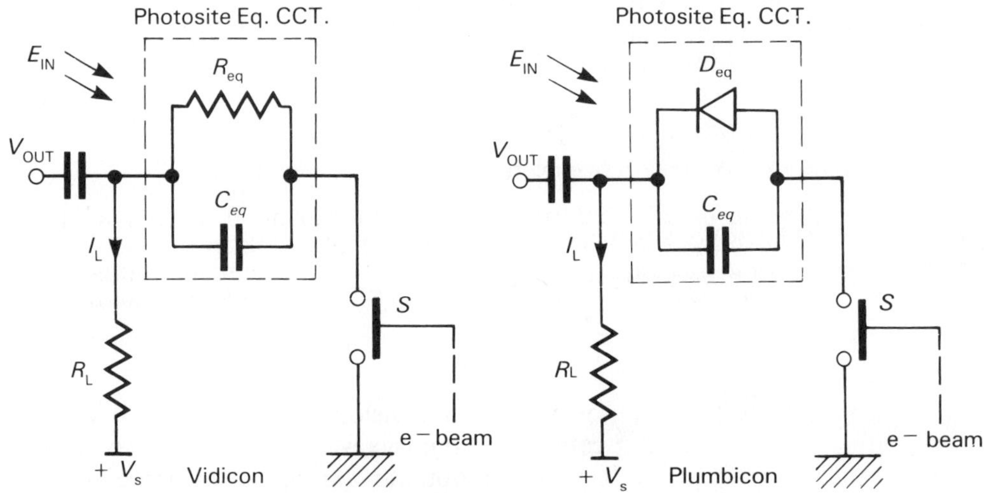

Figure 3.13 Vidicon and Plumbicon principle of operation (after West and Hill)

as the electron beam scans across the target area it 'closes' the switch S on each photosite (i.e. it connects it to the cathode which is at ground potential) thereby charging its equivalent capacitor C_{eq} to a specific voltage V_1. While the beam is scanning another area of the target (i.e. switch S is open), the capacitor discharges through its equivalent parallel component and attains a new, lower voltage V_2.

As the electron beam scans across the target area again, it 'recharges' the photosites' capacitors back to V_1, at a rate I_c proportional to the voltage difference (both the capacitor value C_{eq} and the time interval Δt are in fact constants), as shown in eqn (3.8):

$$I_c = C_{eq} \cdot \frac{dV}{dt} = C_{eq} \cdot (V_1 - V_2)/\Delta t = K \cdot (V_1 - V_2). \tag{3.8}$$

The value of the parallel component, however, is light dependent which means that the capacitor 'recharge' current I_c is proportional to the incident irradiation E_{in}: by monitoring the value of I_c during the scanning period we therefore have a waveform V_{out} proportional to the photosites illumination levels (i.e. to the scene focused on to the target area), as required.

This is summarized in eqn's (3.9) and (3.10):

$$I_c = K \cdot E_{in} \tag{3.9}$$

$$I_L = I_c + I_{dc} = K \cdot E_{in} + I_{dc}, \tag{3.10}$$

where I_{dc} is the dark current, that is the current flowing through the photosite with zero illumination.

Figure 3.13 shows that in a Vidicon the parallel component is equivalent to a Light Dependent Resistor (LDR) and its output characteristic is therefore non-linear (see Section 3.2), the non-linearity being created by the gamma factor γ (sometimes referred to as gamma distortion) shown in eqn (3.11):

$$\left.\begin{aligned} I_L &= [K \cdot (E_{in})^{\gamma}] + I_{dc} \quad (3.11) \\ V_{out} &= R_L \cdot I_L \rightarrow \text{Non-linear} \quad (3.12) \end{aligned}\right\} \text{Vidicon}$$

The Plumbicon parallel component, on the other hand, is equivalent to a reverse biased photodiode which has a linear characteristic (see Section 3.3), as illustrated by eqns (3.13) and (3.14)

$$\left.\begin{aligned} I_L &= [K \cdot E_{in}] + I_{dc} \quad (3.13) \\ V_{out} &= R_L \cdot I_L \rightarrow \text{Linear} \quad (3.14) \end{aligned}\right\} \text{Plumbicon}$$

A comparison between the vacuum television camera types (J. Loebl, 1985) shows that the Vidicon can handle the largest input dynamic range and is therefore less susceptible to image 'blooming' (an apparent defocusing of the bright regions of the television image), it also has the lowest unit cost, the lowest sensitivity (that is it requires a higher minimum input light level) and a less linear input–output relationship due to the gamma distorsion.

The Plumbicon, by contrast, has the lowest image lag (it only takes 2 or 3 scans to 'forget' the last image compared to 4 or 5 for the Vidicon), a good sensitivity and a linear input–output response but does exhibit a high unit cost, caused by the necessity to maintain the lead oxide air-free during the entire manufacturing process. Both the Vidicon and the Plumbicon have a wavelength response similar to that of the human eye, unlike those of the Newvicon and Chalnicon which are far more red and infra-red sensitive (Section 3.5.3).

Newvicons and Chalnicons, on the other hand, feature the highest sensitivity (they will give a useful image in a semi-darkened room) at a unit cost similar to that of the Vidicon.

3.5.2 Solid state optical array transducers

Solid state optical array transducers have been derived mainly from the silicon memories technology, indeed their development has been possible only because of the great strides that microelectronics has made during the 70s and 80s. As a result solid state cameras have a good geometric stability within a flexible configuration. Solid state cameras, in fact, unlike vacuum types, can have either a 2-D photosite array (area scan cameras) or a 1-D photosite array (linescan cameras), the latter geometric configuration having the advantage of using a much smaller silicon area thus producing a higher manufacturing yield and a lower unit cost.

The solid state camera can be based on 4 different photosite arrays:

(i) the Photodiode array

(ii) the Charge Coupled Devices (CCD)
(iii) the Charge Injection Devices (CID)
(iv) the Dynamic Random Access memory (DRAM) used in the optical mode

3.5.2.1 *Photodiode array*

The elementary picture element (or photosite) for this device is the photodiode, that is a light sensitive, reverse biased P–N junction (see Section 3.3 for further details). The operating principle is illustrated in Figure 3.14:

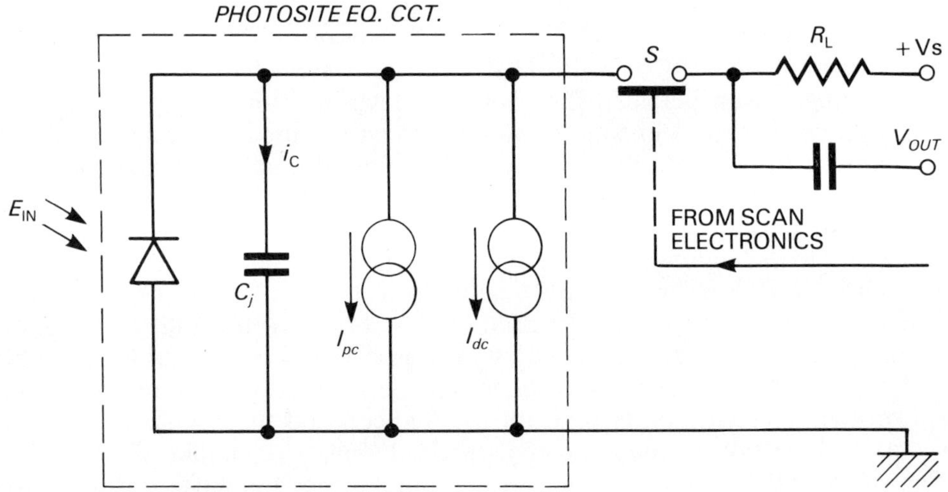

Figure 3.14 Principle of operation of photodiode array

while switch S is closed a charge q is stored in the junction capacitance C_j as given by eqn (3.15):

$$q = C_j \cdot V_s = \int i_c \cdot dt \tag{3.15}$$

When the switch S is open the charge is reduced at a rate proportional to the incident radiation (the photocurrent I_{pc}) and the thermally generated hole-electron pairs (the dark current I_{dc}). For modern, high quality P–N junction (where $I_{pc} \gg I_{dc}$) the discharge time constant is of the order of several seconds for zero incident radiation and in the milliseconds region for typical irradiation levels (Hill, 1982): the charge stored within each photosite P–N junction is therefore inversely proportional to the incident irradiation level. Monitoring the rate of change of this charge (i.e. the capacitor current i_c) while scanning the photosites provides a voltage V_{out} proportional to the grey scale image of the scene focussed on the array, as required. This technology favours uniformity and stability of response but does allow a lower array resolution than other solid state cameras in view of the more complex 'on chip' support electronics required.

The most common geometry for industrial photodiode arrays is in fact the linear array (or linescan) camera. This is due, as previously mentioned, to the technological difficulties involved with making large 2-D photodiode arrays and their relative high cost.

Linescan cameras offer a simple, low cost method of detecting edge position which, in turn, can be used to measure object size parameters such as width or length (IPL, 1984; Electronic Automation, 1985; Hollingum, 1984; Drews *et al.*, 1986), as shown in Figure 3.15 (for further information on how to separate the object image from the background and obtain its edges positions, please refer to Chapter 8 on image processing).

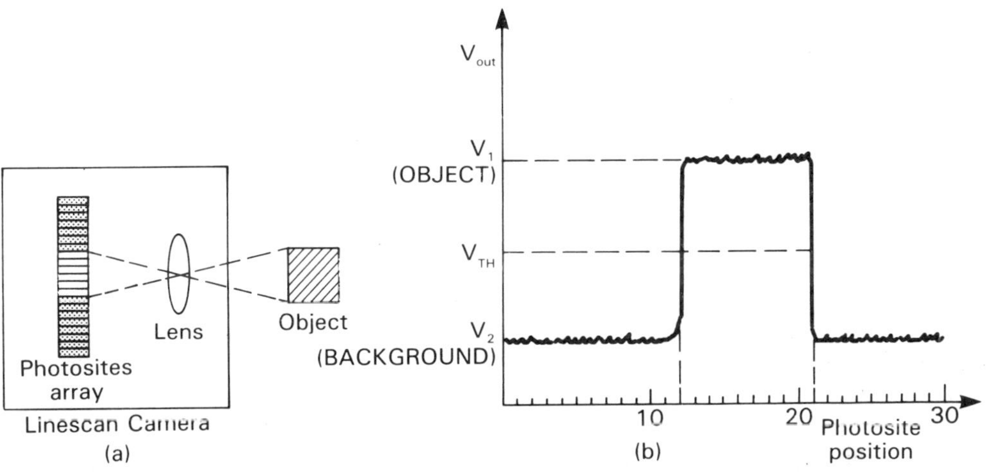

Figure 3.15 Operating principle of linescan camera

A device which uses a linear photodiode array to measure object width based on this technique is shown in Figure 3.16(a). This device can be used in arc welding path control vision systems where the metal channel width and its position with respect to the welding torch need to be measured in real time in the presence of severe electrical and optical interference.

It can also be seen from Figure 3.16 that if the lenses field of view is fixed and the light source is capable of producing a sharp dot on the measurand surface irrespective of its distance then the same arrangement can be used for measuring range. This is, in fact, the principle of operation of some single point rangefinder sensors.

These devices use laser diodes to produce the required collimated light beam and are employed in automated manufacturing applications where non-contact range measurement is essential, as is the case in the food and

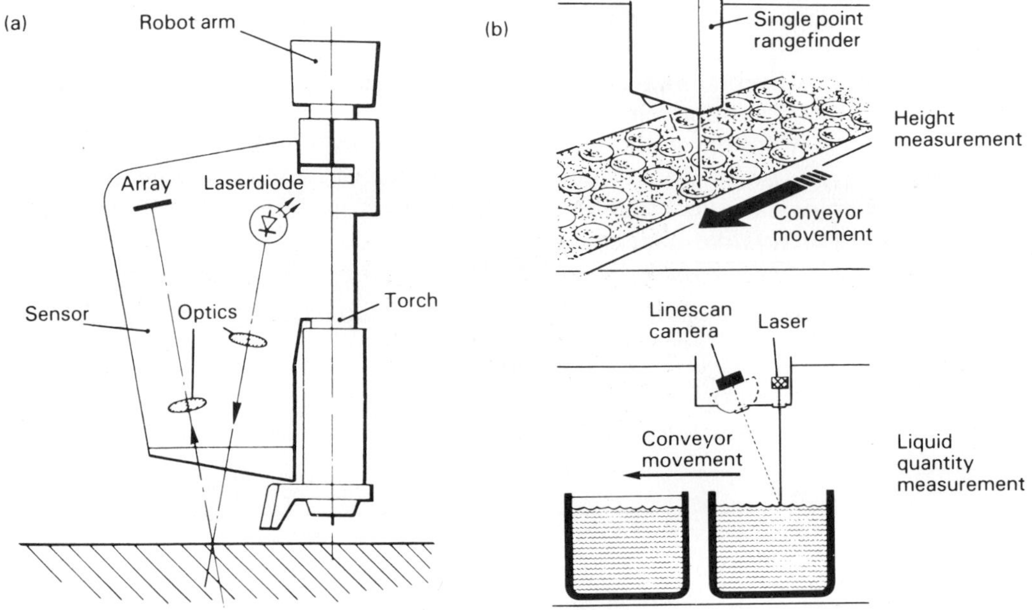

Figure 3.16 (a) Example of vision sensor based on a linear array solid state transducer (after Drews *et al.*, courtesy of *Robotics*); (b) Non-contact inspection in the food processing industry (courtesy of Electronic Automation)

confectionery industry (Electronic Automation, 1984), an example of which is shown in Figure 3.16(b).

3.5.2.2 *Charge Coupled Devices* (*CCD*)

The elementary picture element (photosite) for these devices is based on the Metal Oxide Semiconductor (MOS) capacitor on either a p-type or an n-type substrate, as shown in Figure 3.17(a) for an n-type substrate device

The negative potential applied to the photosite electrode creates a depletion region (known as a 'potential well') within the substrate. The size of the potential well depends on the electrode shape for width and length and the value of the applied voltage for depth. When the photosite is optically irradiated the incident photons generate hole-electron pairs by virtue of the internal photoeffect (as described in Section 3.1). The mobile carriers (electrons in this case) are repelled by the applied voltage and leave behind fixed charges whose number is proportional to the incident radiation E_{in}, this is therefore equivalent to generating a photocurrent I_p.

Any hole-electron pairs thermally generated within the potential well are also separated by the applied voltage and provide additional charges which, being indistinguishable from the photon generated ones, are equivalent to a 'dark' current I_{dc}.

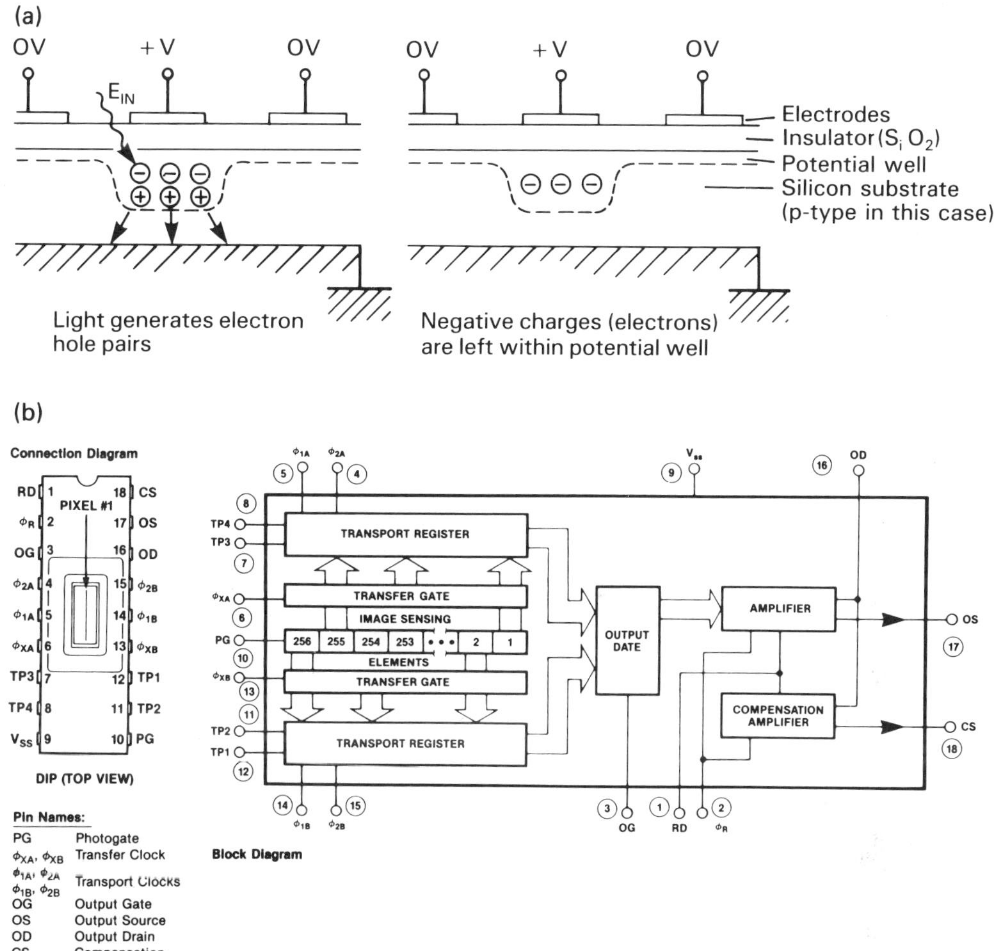

Figure 3.17 (a) Charge Coupled Device principle of operation (after West and Hill, 1982); (b) CCD111 linescan camera (courtesy of Fairchild, 1984)

After a chosen time interval, which is generally referred to as the light 'integration time', the charge thus accumulated in the potential well is transferred via gates to a charge coupled shift register (hence the device name) which transports it to a charge sensitive video amplifier, as shown in Figure 3.17(b). By repeating the process for each photosite (i.e. by scanning the photosite array) the output of the video amplifier will be an electrical signal proportional to the scene focussed on the array, as required.

CCD solid state cameras can be based on either a 2-D photosite array (area scan cameras) or a 1-D array (linescan cameras). They are geometrically very stable and accurate devices with a good input dynamic range and are therefore becoming increasingly popular in the fields of robotics and metrology. Their relatively high unit cost, however, has tended to restrict their use to automated inspection applications, such as the GIOTTO European space probe or the British Central Electricity Generating Board nuclear plant maintenance robot. Industrial applications of the CCD camera are, however, growing rapidly as advancements in microelectronics technology increase the CCD optical resolution and help to reduce its unit cost.

3.5.2.3 *Charge Injection Devices* (*CID*)

These devices are also based on the MOS capacitor and indeed employ the internal photoeffect in the same way as the Charge Coupled Device (CCD).

The essential difference lies in the charge 'reading' operation. Unlike the CCD, in fact, the CID do not transfer the charge along to a video amplifier but 'inject' it (hence their name) into the substrate by reducing the electrode voltage to zero (i.e. by removing the potential well) thus causing a photocurrent I_p to flow, as shown in Figure 3.18:

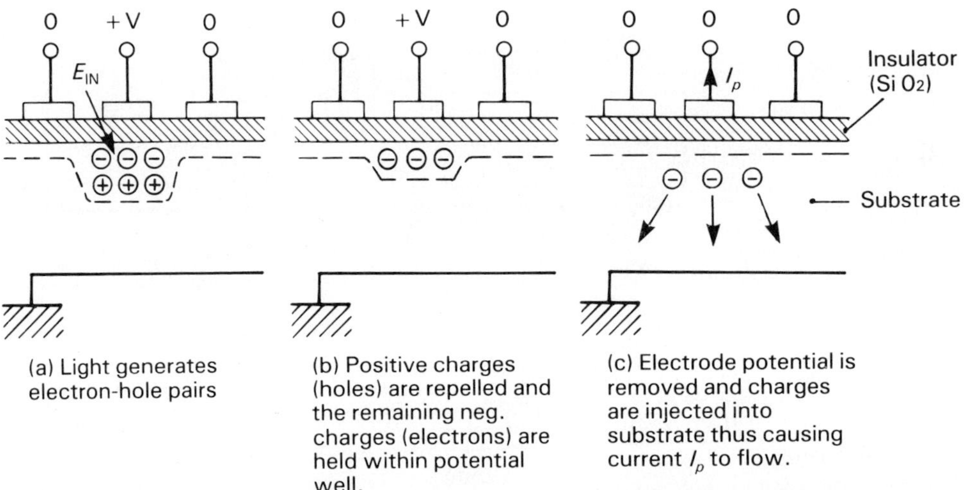

Figure 3.18 Charge Injection Devices principle of operation (after West and Hill)

The photocurrent I_p is, in effect, the MOS capacitor displacement current, as shown in eqn (3.16):

$$I_p = C \cdot \frac{dV}{dt} = C \cdot \frac{V}{\Delta_t} = \frac{q}{\Delta_t} = K \cdot q. \qquad (3.16)$$

Since the MOS capacitor value C and the time interval Δt (necessary to take the electrode voltage from V to 0) are both constants the height of this 'injected' photocurrent pulse is proportional to the stored charge q and therefore to the incident irradiation E_{in}:

$$I_{p(\max)} = K \cdot E_{in}. \tag{3.17}$$

By monitoring the displacement current peak value during the photosites scan an analogue signal V_{out} proportional to the image intensity is obtained, as required. CID solid state cameras have similar features to CCD ones, being based on the same MOS technology, namely stable and accurate geometric configuration with a good input dynamic range and a high unit cost. CID cameras are, however, more tolerant of excess light input levels and suffer from less cross-talk than CCD cameras because of the absence of the charge transfer mechanism.

3.5.2.4 *Dynamic Random Access Memories* (*DRAM*)

These optical transducers make use of an ordinary DRAM chip whose ceramic package has had the lid removed so as to allow light to fall on to the memory cells, as shown in Figure 3.19:

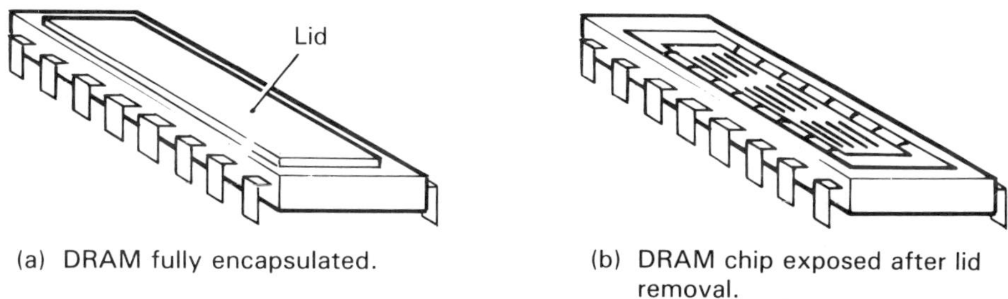

(a) DRAM fully encapsulated.

(b) DRAM chip exposed after lid removal.

Figure 3.19 DRAM camera

These devices therefore work by 'memorizing' the light pattern focused on the memory cells plane. The principle of operation includes 3 basic steps:

(i) the photosites array is 'coated' with electrical charges, that is all the memory cells are filled with logic '1's
(ii) the memory cells are then disconnected from the data bus whilst being exposed to light which, by creating electron-hole pairs (photo-electric effect) beneath the SiO_2 gates, allows each cell to slowly discharge at a rate dependent on the amount of light falling on it.
(iii) after a time interval, which by analogy with other cameras we shall call 'exposure time', the memory array is read, thus providing binary data corresponding to the light pattern falling on the array during the exposure time.

The features of this simple, yet very effective area scan camera, depend, predictably, on the MOS memory technology used. Memory cell design, in fact, aims to reduce the cell size, the stray capacitances and the signal path lengths in order to obtain the highest cell density with the lowest access time.

This may lead to a physical cell layout unsuitable for use as a light transducer and therefore not all DRAM chips available on the market are suitable for use as a solid state camera. Even those that have been commercially used require some image pre-processing in order to produce an acceptable image, such as the removal of the black line caused by the memory cells power supply rail. Two other drawbacks of these devices are the limited input dynamic range (and the relative long exposure times required) and the inherent binary nature of the image (since each cell output can only be a logic 1 or 0).

A grey scale image using this device can, however, be produced at the expense of image acquisition time by overlaying several binary images taken with different exposure times. DRAM cameras are inexpensive in comparison with other optical array transducers because the optical array transducers used (the DRAM memory chip) is manufactured in large numbers for the computing industry. This feature has produced wide acceptance of the DRAM camera in low cost vision systems, such as those used on educational robots and in light industrial automation (Lambert, 1984; Owen, 1982).

3.5.3 Comparison of vacuum to solid state camera technologies

To carry out a comprehensive comparison of the vacuum versus solid state technology one would also need to include a comparison of the specifications for the various cameras available on the market, since there can be considerable performance variations between similar cameras made by different manufacturers. An in-depth comparison can therefore only be carried out by referring to the manufacturers data sheets and data manuals, some of which have been listed as further reading material at the end of the chapter. General guidelines from this comparison can, however, be drawn and are reported in this paragraph.

Vacuum cameras can suffer from *geometric distortion* due to the non-linearities in the electromagnetic scanning; solid state cameras exhibit a better geometric stability and linearity than vacuum ones and are therefore better suited for optical dimensional measurements, such as in the case of automatic arc welding control [Meta Machines 1985, Edling 1986].

The *wavelength response* of the two technologies is shown in Figure 3.20(a) and 3.20(b) from which it can be seen that solid state cameras are more sensitive to longer wavelength (such as infrared light) than Newvicons and Chalnicons but much more so than the most popular vacuum cameras, namely the Vidicon and the Plumbicon. I.R. sensitivity can be a drawback in terms of image focusing and resolution (since the camera optics are affected by the wavelength), and might therefore require the use of an I.R.filter, but can also be an advantage if non-visible illumination is required, such as in

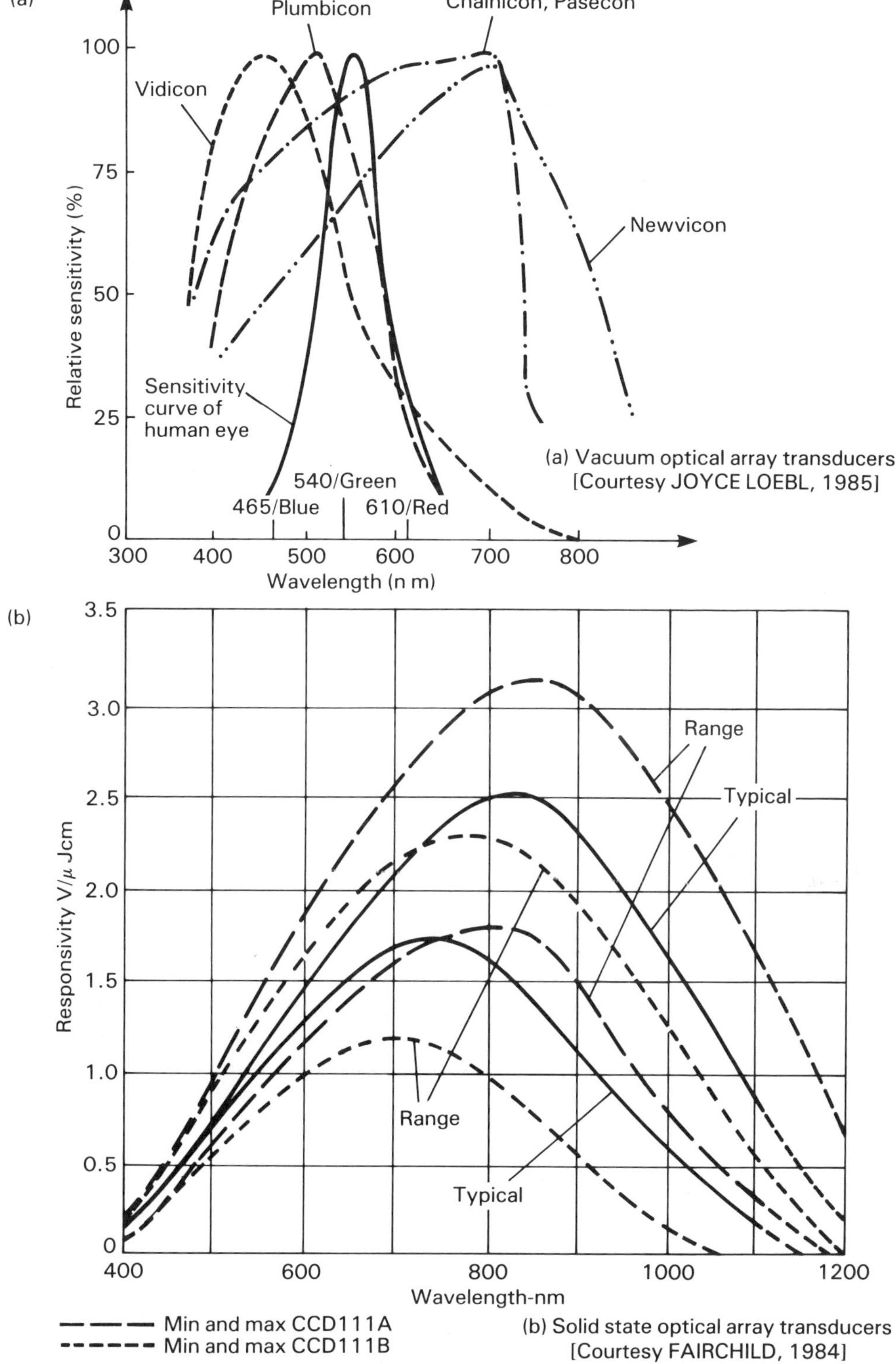

Figure 3.20 Typical spectral response of some vacuum and solid state cameras

the case of imaging objects moving on a conveyer belt where strobed I.R. illumination may need to be used in order to avoid creating unpleasant conditions for factory workers (see Section 6.2).

Solid state cameras have a very low image lag and can therefore be used in higher *speed* applications than vacuum cameras. They also allow a greater *flexibility* in vision system design since solid state cameras come in different geometries (1-D and 2-D array devices are available) and with different resolutions (64 to 4096 pixels is a typically resolution range for linescan cameras). These features can be used to help improve specific vision system parameters, such as system cost or image processing complexity and speed, as demonstrated in a food processing plant where the use of medium resolution, linescan cameras enables the high speed inspection of corn cobs (Hollingum, 1984).

The smaller *size* and more rugged construction of solid state cameras, afforded by the absence of a vacuum envelope, also allows a wider range of applications compared to vacuum cameras, such as in the case of eye-in-hand vision sensors (Pugh, 1982; Loughlin and Hudson, 1982; Van de Stadt, 1985).

Solid state cameras have a higher *sensitivity* than the Vidicon but, typically, a lower one than the Newvicon and Chalnicon. Solid state cameras do, however, have a less homogeneous sensitivity across the photosite due to the photosite physical size tolerances. A typical CCD photosite is shown in Figure 3.21; its dimensional tolerances are generated mainly by inaccuracies in the electron beam lithographic masks used during the device manufacture and yield a variable photosite active area which, in turn, produce variations in the individual photosite sensitivity: in the CCD case, for instance, the worst case photosite size error is $\pm 0.065\ \mu m$ for 1024 pixels at 13 μm pitch (Fairchild, 1984) which, as shown by eqn 3.18, produces a photosite sensitivity error ΔS_p of $\pm 0.5\%$:

$$\Delta S_p = \frac{\Delta L_p}{L_p} \cdot 100 = \frac{0.065}{13} \cdot 100 = 0.5\% \tag{3.18}$$

Vidicons, on the other hand, have a variable sensitivity which can be

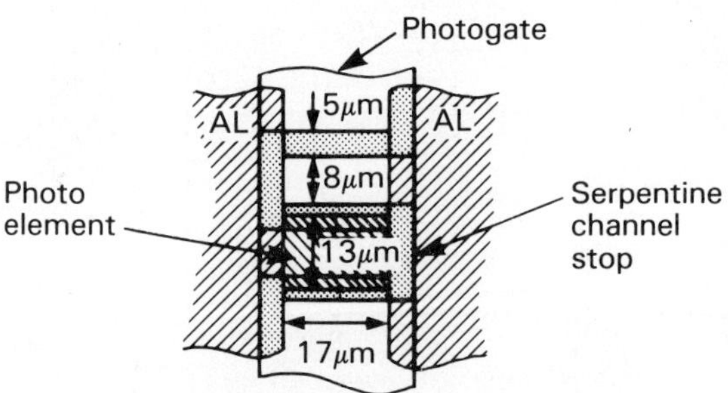

Figure 3.21 Typical photoelement (or photosite) dimensions for the CCD111 linescan camera (courtesy of Fairchild, 1984)

changed by adjusting the target voltage (other devices have a fixed sensitivity) and a higher *dynamic range*; this latter, however, is afforded largely by the gamma factor non-linearity and the lower sensitivity and might not always be an advantage.

Electrical *noise* tends to be higher in vacuum devices because of the thermoionic, and therefore random, nature of the electron emission from the vacuum cameras cathode. Some noise is, however, also generated within CCD cameras as a consequence of the cross-talk that takes place between contigous photosites and during the charge transfers.

Ultimately *cost* may also play a part in the device selection even though the camera cost is typically only a small proportion of the overall vision system cost. The Vidicon is generally the cheapest area scan camera on the market (with the exception of DRAM cameras which however have a worse performance and do not find large use in the industrial robotics field) due mainly to the large number of devices manufactured for the television market, whereas the Plumbicon tends to be the most expensive device among those described. It should be noted that solid state linescan cameras are, in fact, cheaper than Vidicons but that a direct price comparison is somewhat meaningless in view of their different array geometries and specifications.

3.6 Conclusions

Optical transducers are based on the photoelectric effect and vary in cost and size from the subminiature photodiode to the wide angle photomultiplier. The most popular optical transducers in the robotic field are usually spectrally matched Light Emitting Diode/Photodiode combinations, used mostly in proximity detectors, and the solid state array cameras, used for scene imaging and subsequent processing.

3.7 Revision questions

(a) The circuit shown in Figure 3.a uses the photoresistor whose transfer characteristic was shown in Figure 3.3. Calculate a suitable set of

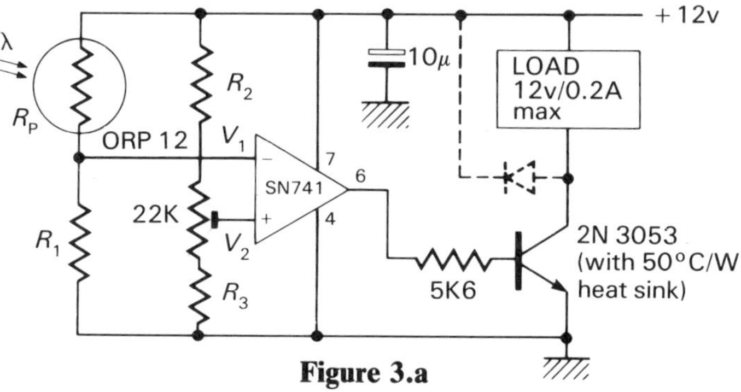

Figure 3.a

component values to turn the robot gripper on whenever an object appears between its fingers.

(b) The photodiode in the circuit shown in Figure 3.b has a conversion efficiency of 0.5 A/W, an active area of 1 mm^2 and a stray capacitance of 10 pF at -10 V. Calculate the value of R_L to yield a break frequency of 50 MHz and the optical flux on the photodiode plane to give an output voltage of -1 V. Assume the op. amp. is ideal.

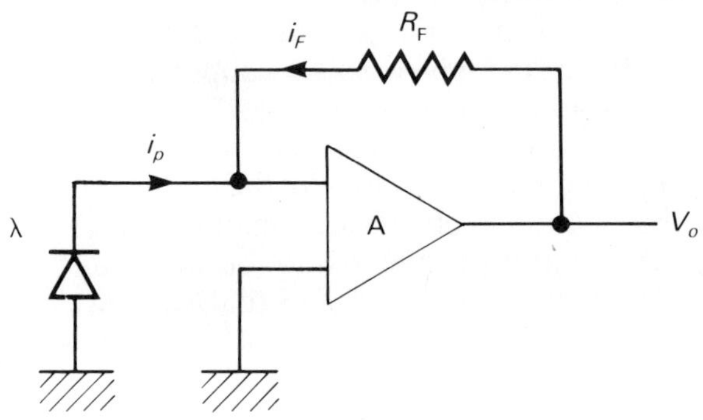

Figure 3.b

(c) Carry out a survey on the use of solid state cameras in robotics application of your choice. What conclusions can you draw from this mini-survey?

3.8 Further reading material

1. (1976) *Optoelectronics—Theory and Practice,* Texas Instruments.
2. (1982) *Optoelectronics,* General Electric.
3. (1982) *Photomultiplier Tubes,* Hamamatsu.
4. (1984) *Charge Coupled Devices* (CCD) *Catalog* Fairchild.
5. Joyce Loebl (1985) *Image Analysis*: *Principles and practice.*

Chapter 4

Force transducers

4.1 Overview

Force transducers are required to transform a physical force into an electrical signal. Their uses in the Robotics field have traditionally been confined to specialized, custom-built applications and therefore are limited in number. The advent and development of adaptive control techniques, however, have increased the popularity of force and torque sensors which will allow the next generation of robots to interactively control the gripping action as well as the object movement during the programmed task.

The adaptive control of a complex mechanical structure, such as a robot, requires the measurement of the physical forces produced during the gripper movement and/or object manipulation. These forces can be broken down into two main categories:

- Kinematic forces, such as those exerted on the wrist assembly by the acceleration of the object mass during a move operation.
- Static forces, such as those exerted by the gripper on to the object surface during a manipulation operation.

This latter category has further led to the development of force sensors used within the gripper and which, because of their human analogy, are called *tactile sensors*. These devices will be described in detail in Chapter 7.

4.2 Force measurement generalities

Force transducers perform the mechanical-to-electrical energy conversion by measuring the physical deformation that the force in question has produced

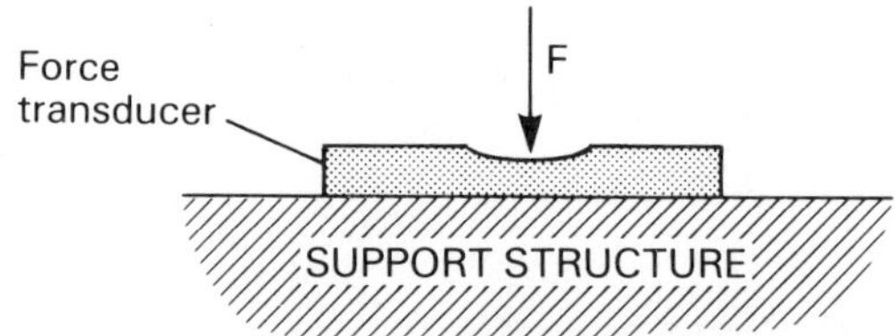

Figure 4.1 Direct measurement, force deforms transducer only

on the transducer. This deformation can be caused directly, as shown by Figure 4.1, or more commonly transmitted to the transducer via a support material, as shown in Figure 4.2. The choice and shape of the support material in the latter case greatly affects the performance of the force transducer, in particular its dynamic measurement range. The long-term stability of the support material and its adhesion to the transducer itself also affect the measurement and require consideration when designing a force transducer system.

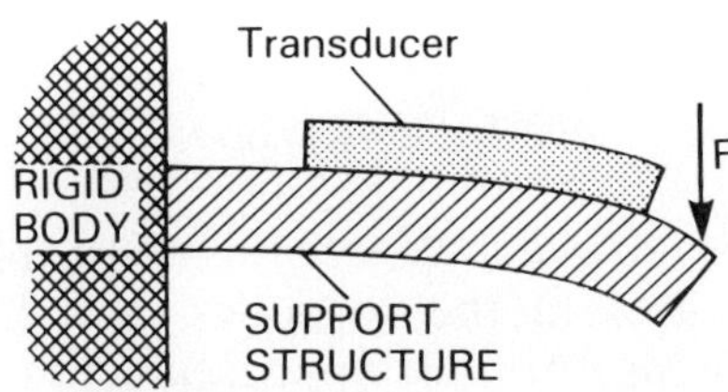

Figure 4.2 Indirect measurement, force deforms both support and transducer

The direct measurement method shown in Figure 4.1 is used mainly in tactile sensors and therefore will be described in Chapter 7. The indirect method of force measurement via a support material will be assumed throughout the rest of this chapter, therefore a brief description of the influence of the supporting structure on the force measurement is required. There are three main types of force which the transducer needs to measure: traction/compression, bending and twisting force.

4.2.1 Traction/compression force measurement

This force, as illustrated in Figure 4.3, is applied longitudinally to a support structure such as to cause an increase (for traction) or decrease (for compression) of its length $\pm\Delta L$; an example of a structure under traction stress is one of the limbs of the robot in the vertical position. The physical deformation ΔL produces a change in the transducer unstressed parameter value V_t in accordance with a constant of proportionality K:

$$\Delta V_t = K\Delta L(V_t)$$

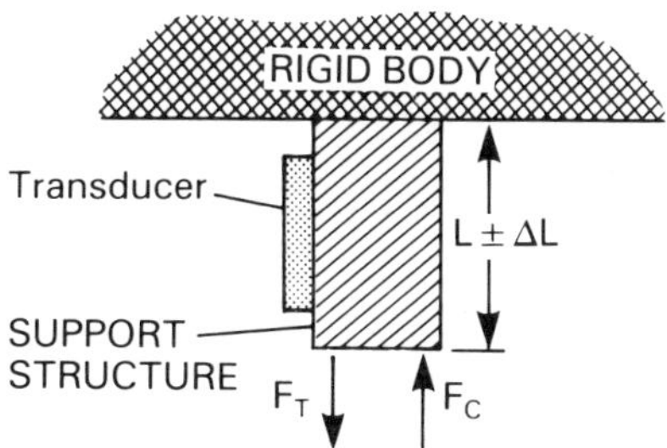

Figure 4.3 Traction/compression force measurement principle

The unit change in the transducer parameter value ($\Delta V_t/V_t$) is the quantity usually measured. The force applied can be obtained if the supporting structure cross-sectional area A, its Young's modulus E and the transducer's sensitivity G_f are known, as shown by eqn (4.1), Young's modulus being the ratio E = (stress/strain)

$$F = \frac{EA}{G_f}\left(\frac{\Delta V_t}{V_t}\right) \tag{4.1}$$

The unit change in the transducer parameter value ($\Delta V_t/V_t$) can be measured using a variety of techniques, some of which will be described later in this chapter.

4.2.2 Bending force measurement

A bending force can be defined as that which causes a longitudinal flexion in a support structure, as illustrated in Figure 4.4. An example of structure

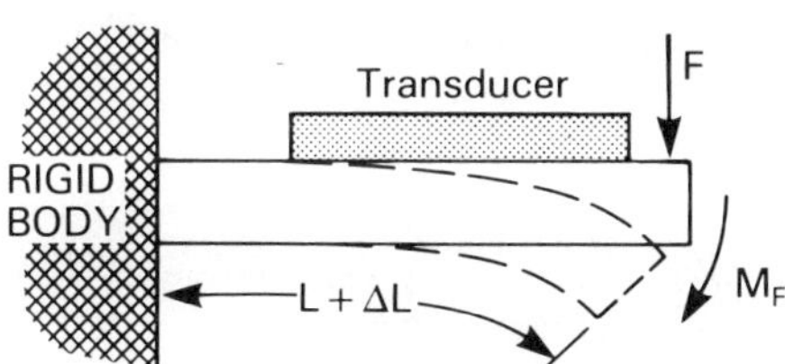

Figure 4.4 Bending force measurement principle

subject to a bending force is a robot limb in a non-vertical position with a mass suspended at its free end (which can be either the object mass or, as in this case, the robot next limb), as illustrated in Figure 4.5. Having measured the unit change in the transducer unstressed parameter ($\Delta V_t/V_t$), the flexing moment M_f can be obtained as shown in eqn (4.2).

$$M_f = (\Delta L)EW = \frac{EW}{G_f}\left(\frac{\Delta V_t}{V_t}\right) \tag{4.2}$$

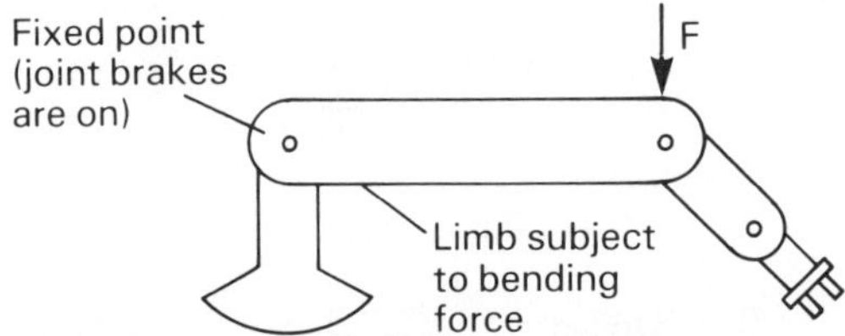

Figure 4.5 Example of robot limb subject to bending force caused by weight of forearm and gripper

where W is the stiffness of the support structure to a bending force, E is its Young's modulus and G_f is the transducer sensitivity.

4.2.3 Twisting force measurement

This force, as the name suggests and Figure 4.6 illustrates, is present in all rotating structures, such as the drive shaft of a robot actuator. The measurement of the twisting force on such a shaft may, for instance, be required to monitor the mechanical resistance to the shaft rotation. The twisting moment M_t can be derived as shown in eqn (4.3):

$$M_t = \left(\frac{4S_T I_p}{dG_f}\right)\left(\frac{\Delta V_t}{V_t}\right), \tag{4.3}$$

where S_t is the tangential stiffness of the shaft, d is the shaft diameter and I_p is the moment of polar inertia of the shaft cross-section which, for a cylindrical shaft, is given by $0.1\,d^4$; $(\Delta V_t/V_t)$ still represents the measurement of the unit change in the transducer parameter value.

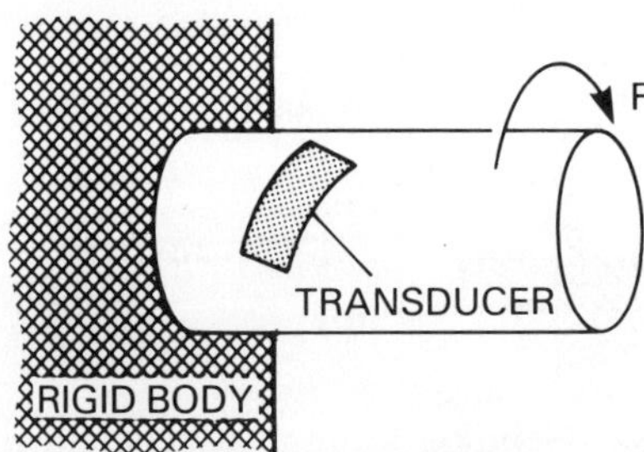

Figure 4.6 Twisting force measurement principle

4.3.4 Force transducer generalities

There are three main types of force transducers: resistive, semiconductor (also sometimes called piezoresistive) and non-resistive. The last, which includes capacitive and fibre-optic types, do not find many applications within a robot structure but can be used in some of the peripheral systems and have therefore been included in this chapter. Resistive and semiconductor transducers are generally referred to as *strain gauges*.

4.3 Strain gauges

The strain gauge is a transducer used for the measurement of dimensional variations within mechanical systems. Its principle of operation is based on the variation in electric resistance within a conductor with respect to a change in its physical dimensions, usually length.

An indication of the sensitivity of a strain gauge is therefore the ratio between the unit change in the electrical resistance and the unit change in the physical length, as shown in eqn (4.4); this is known generally as the *gauge factor,* which is denoted in this book by the symbol G_f.

$$G_f = \frac{\Delta R/R}{\Delta L/L}. \tag{4.4}$$

Initially, strain gauges were constructed with a length of metal wire and were consequently somewhat bulky. These days strain gauges are divided in to two main groups: metal (resistive) and semiconductor (piezoresistive).

Metal strain gauges are made of a metal film pattern, typically only a few micrometres thick, bonded to a flexible and electrically insulating backing material such as polyester. This construction (Figure 4.7), is obtained with techniques similar to those employed in the microelectronics industry to produce film resistors and therefore benefits from the same high reliability and low tolerance values.

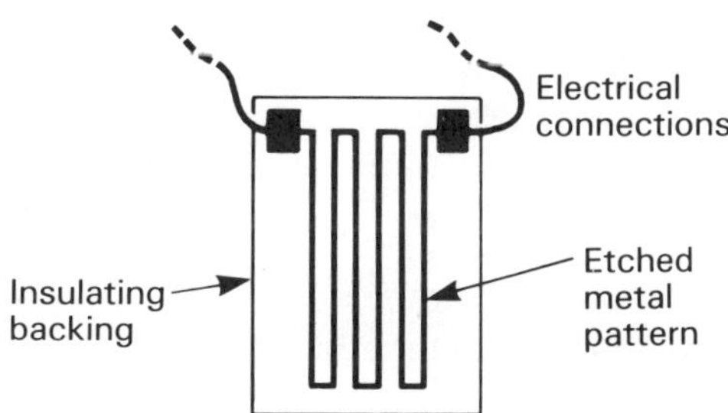

Figure 4.7 Metal strain gauge (resistive)

The gauge factor of metal strain gauges is given by eqn (4.5), where μ is the Poisson ratio, which relates the increase in the length of a solid object under the effect of a traction force to the reduction of its diameter.

$$G_{f(\text{metal})} = \frac{\Delta R/R}{\Delta L/L} = 1 + 2\,\mu \tag{4.5}$$

Table 4.1 lists typical gauge factors for metal strain gauges.

Table 4.1 Typical strain gauge metallic compositions and relative gauge factors.

Material	*Composition* (%)	*Gauge factor*
Alloys		
Nichrome	Ni80, Cr20	+2.0
Manganin	Ni4, Mn12, Cu84	+0.5
Cromel-C	Ni64, Fe25, Cr11	+2.5
Isoelasic	Ni36, Cr8 Fe55.5, Mo0.5	+3.5
Pure metals		
Nickel	—	−12.0
Aluminium	—	
Platinum	—	+4.8

Semiconductor strain gauges, sometimes referred to as piezoresistive strain gauges, are based on thin layers of doped silicon or germanium manufactured into rectangular sections, as shown in Figure 4.8. The gauge factor for these transducers is given by:

$$G_f = \frac{\Delta R/R}{\Delta L/L} = 1 + 2\mu + p_c E \qquad (4.6)$$

where μ is Poisson's ratio, p_c is the transducer longitudinal piezoelectric coefficient and E is Young' modulus.

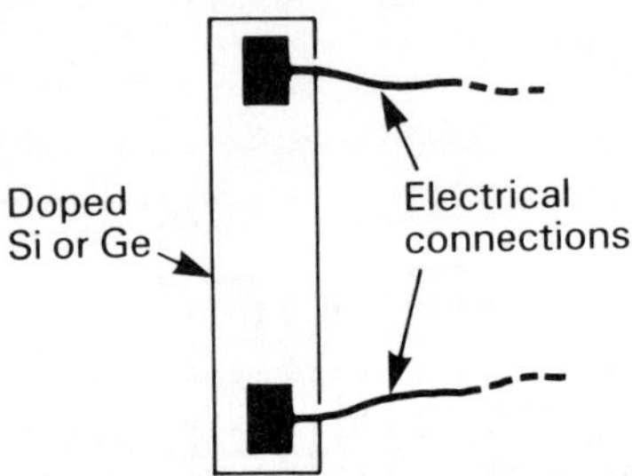

Figure 4.8 Semiconductor strain gauge (piezoresistive)

Table 4.2 Semiconductor strain gauges.

Material	*Resistivity* (Ω cm)	G_f
Ge(n)	1.5	−147
Ge(p)	1.1	+101.5
Si(n)	11.7	−9.3
Si(p)	7.8	+175

Because of the addition of the $p_c E$ term these gauges have a higher gauge factor, that is a higher sensitivity, than metal ones. This is shown in a quantitative format in Table 4.2.
Semiconductor strain gauges can thus measure lower mechanical strains and therefore lower forces than metal ones by approximately two orders of magnitude. However, they have a lower temperature range and lower mechanical strength than metal ones and are therefore limited in their use to high sensitivity, low temperature and low dynamic range applications.

4.3.1 Resistance measurement

As illustrated previously in Figure 4.3 a strain gauge fixed to a structure under traction will exhibit an increase in length and, as shown in eqn (4.1), a change in the output parameter ΔV_t. For strain gauges this latter represents a change in resistance ΔR; eqn (4.1) can therefore be re-written as eqn (4.7) in which the only unknown is ΔR, since G_f and R (gauge factor and unstressed gauge resistance) can be obtained from the transducer manufacturer's data sheet, while E and A (Young's modulus and the cross-sectional area) are provided by the support structure design.

$$F = \frac{EA}{G_f}\left(\frac{\Delta R}{R}\right) = K_T \Delta R \tag{4.7}$$

A measurement of the strain gauge resistance change ΔR is therefore sufficient to obtain the value of the force acting on the structure. Although several different methods of resistance measurement could be used for this purpose, the most popular one, in view of its sensitivity and flexibility of operation, is the Wheatstonc Bridge which is the only method described in this book.

In the basic Wheatstone Bridge circuit the unknown resistance R_x is measured by comparison with three known values, as shown in Figure 4.9; this method of measurement is based on varying the known resistance values until the bridge is balanced, that is the condition $V_0 = 0$ is obtained, at which point the unknown resistance can be calculated using the bridge balance

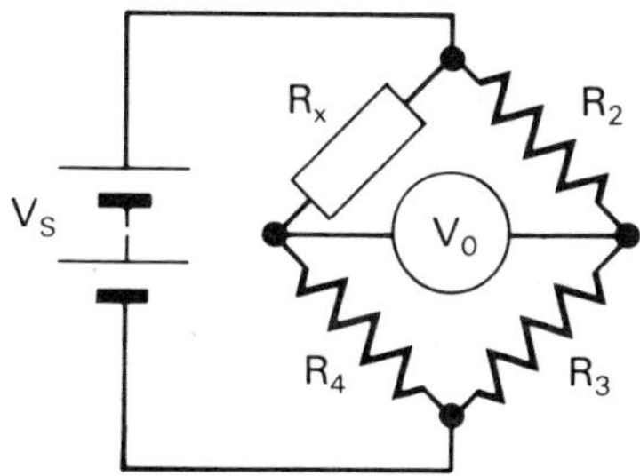

Figure 4.9 Basic Wheatstone bridge circuit

relationship:

$$\frac{R_x}{R_4} = \frac{R_2}{R_3}$$

Any variations in R_x subsequently produce a non-zero output voltage V_0 proportional to the resistance change ΔR_x; the value of V_0 therefore provides a measure of the force acting on the structure under test.

A practical bridge circuit is composed of four equal resistances, that is the three fixed resistances are chosen to be of the same value as the strain gauge unstressed resistance, plus a bridge balancing network. This latter can be of three basic types, as shown in Figure 4.10. For these types of circuit where $R_x(\text{unstressed}) = R_1 = R_2 = R_3$, the output voltage variation ΔV_0 is

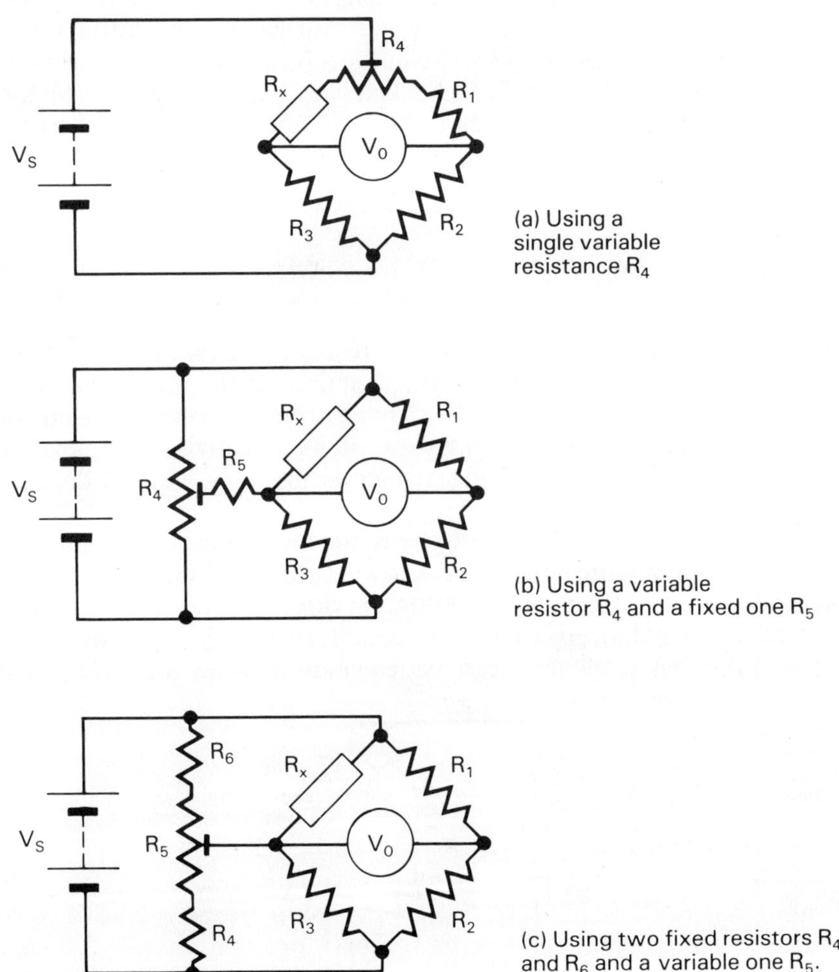

Figure 4.10 Three different methods of zeroing a Wheatstone bridge

given by:

$$\Delta V_0 = \frac{V_s \Delta R}{(4R + 2\Delta R)} = \frac{V_s}{2R} \cdot \frac{\Delta R}{(2 + \Delta R/R)} \tag{4.8}$$

This shows that the $\Delta V_0/\Delta R$ relationship is only linear for small resistance changes, that is small $\Delta R/R$. This is in fact the case in practice and we can therefore approximate eqn (4.8) to:

$$\Delta V_0 \approx \frac{V_s}{4} \cdot \frac{\Delta R}{R} = \frac{V_s \cdot G_f}{4} \cdot \frac{\Delta L}{L} \tag{4.9}$$

where L is the unstressed length of the strain gauge and ΔL its increment under stress. The ratio $\Delta L/L$ is a measurement of the mechanical strain and can therefore be inserted in Young's modulus formula $E = \text{stress/strain}$:

$$\text{Stress} = E \cdot \text{strain} = \frac{E \Delta L}{L} \tag{4.10}$$

The level of the required mechanical stress on the structure under test can thus be measured by the out-of-balance output voltage ΔV_0, given that the bridge supply voltage V_s, Young's modulus for the structure material, E and the force transducer gauge factor, G_f are known. To reduce the Wheatstone bridge temperature dependence a second strain gauge $R_{x(comp)}$ can be inserted in the circuit in place of R_2 and mounted on the structure perpendicular to the first gauge so as to minimize any longitudinal elongation, as shown in Figure 4.11.

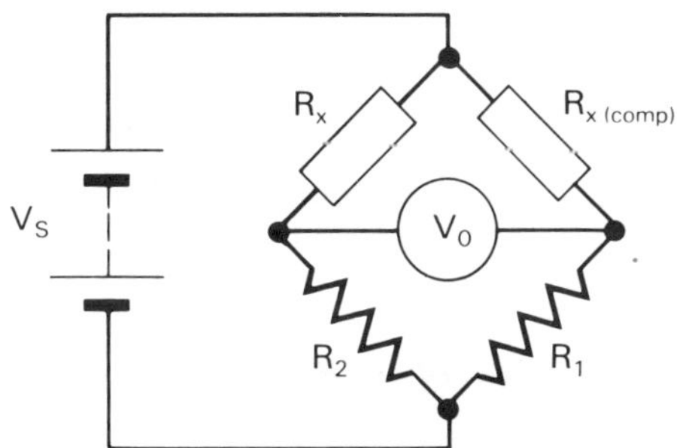

Figure 4.11 Temperature-compensated Wheatstone bridge

To double the bridge sensitivity, another 'active' gauge R_y can be mounted on the side opposite to R_x on the structure under test, in such a way that both gauges are subject to the longitudinal elongation produced by the strain, that is their axes are both parallel to the structure longitudinal axis. To also compensate R_y against temperature variations a fourth gauge $R_{y(comp)}$ is mounted perpendicular to R_y in such a way as to minimize the effect of the strain on its resistance value (Figure 4.12).

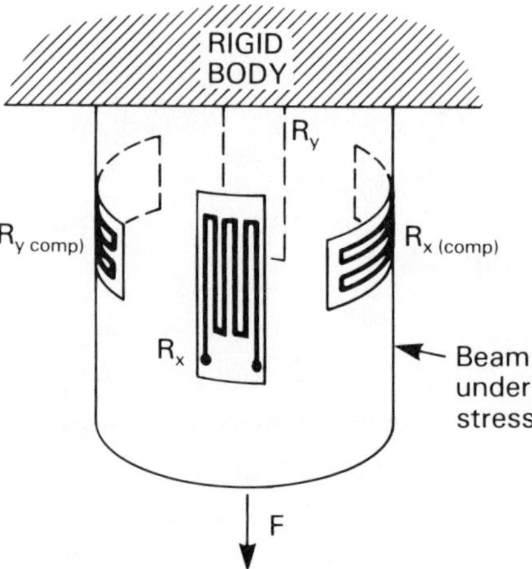

Figure 4.12 Example of traction force measured using four strain gauge in a dual-active temperature-compensated configuration

The overall sensitivity of such bridge is slightly more than double because the compensating gauges also contribute a resistance change $\Delta R_{x(comp)}$ and $\Delta R_{y(comp)}$ due to variations in gauge width. However, Poisson's ratio shows that an increase in width must be followed by a reduction in length, which means that the compensating gauges' resistance changes are negative with respect to R_x and R_y but add to the overall bridge performance because they are connected electrically on the opposite side of the Wheatstone bridge to R_x and R_y. The overall bridge sensitivity is therefore given by:

$$\Delta V_0 = \frac{V_s G_f}{} \left(\frac{\Delta R_x}{R_x} + \frac{\Delta R_{x(comp)}}{R_{x(comp)}} \right) \tag{4.11}$$

$$\Delta V_0 \approx \frac{V_s G_f}{2} \left(\frac{\Delta L}{L} \right) (1 + \mu) \tag{4.12}$$

Using eqn (4.10) we can therefore re-write eqn (4.12) as:

$$\Delta V_0 = K_{tot.} \cdot \text{stress} \tag{4.13}$$

where μ is Poisson's ratio and $\Delta L/L$ is the strain. Equation (4.13) shows that the output voltage change ΔV_0 is still proportional to the mechanical stress acting on the structure, as required, but that, for the same stress (that is, the same variations in gauge resistances ΔR), the bridge output voltage ΔV_0 is greater by more than a factor of two when compared with the basic circuit shown in Figure 4.9.

Similar derivations can be obtained for bending and twisting force applications to increase the strain measurement sensitivity and reduce temperature dependence. The physical locations of the gauges in these latter cases are shown in Figures 4.13 and 4.14. Note that for the twisting force

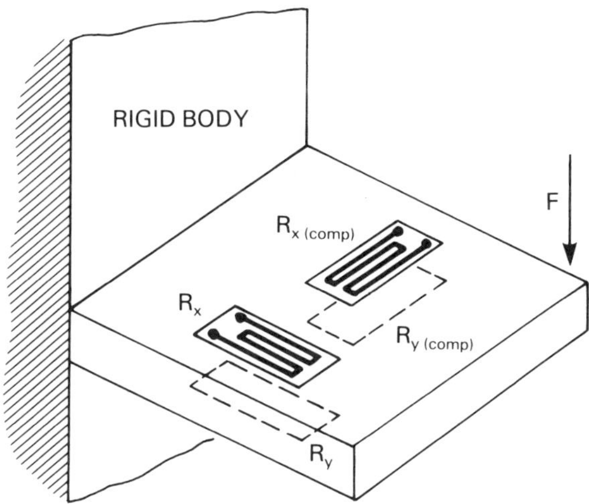

Figure 4.13 Example of bending force measured using four strain gauges in a dual-acting temperature-compensated configuration

case all the gauges are 'active', that is they all receive dimensional variations along their length and there the overall bridge sensitivity is four times that of the single-gauge uncompensated circuit shown in Figure 4.9. Also it should be pointed out that any bending moments acting on the structure will introduce an error since the mounting configuration shown does not compensate for them.

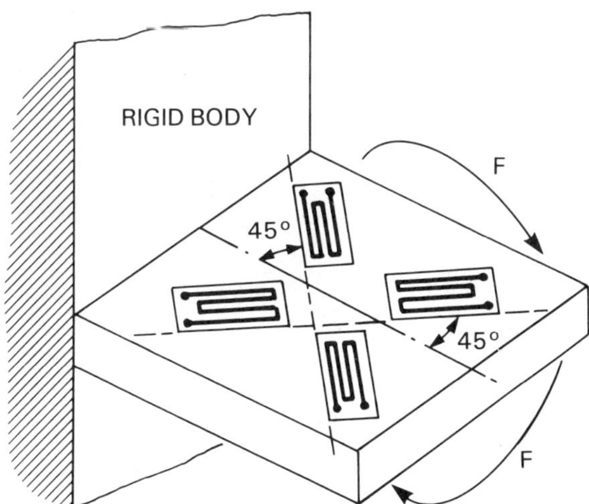

Figure 4.14 Example of twisting force measured using four strain gauges in a temperature-compensated configuration

4.4 Non-resistive transducers

The most widely used non-resistive force transducer is the capacitive type which proeuces a change in capacitance proportional to the pressure exerted on the plates. The principle of operation is similar to that of a compression capacitor used to provide small values of variable capacitance in electrical circuits. This principle is illustrated in Figure 4.15. The variation in

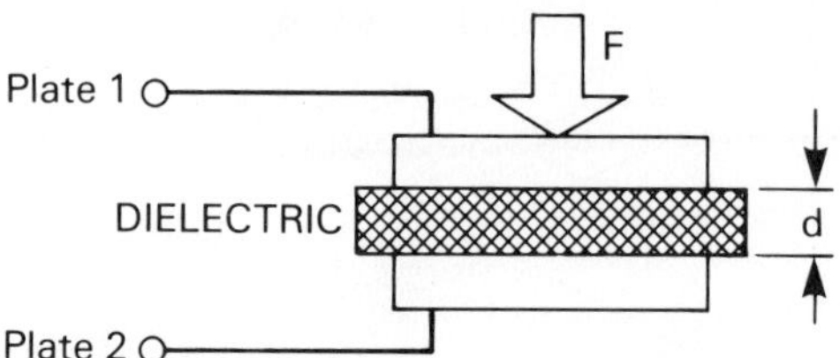

Figure 4.15 Simplified diagram of a compression-type variable capacitor

capacitance ΔC depends on the change in distance, d between the capacitor plates and therefore on the compression characteristics of the dielectric material which, in turn, depends on Young's modulus E, as shown by:

$$C = E\frac{A}{d} \tag{4.14}$$

Differentiating with respect to d and taking small increments we can re-write eqn (4.14) as:

$$\frac{\Delta C}{C} = \frac{d}{\Delta d} = \frac{E}{\text{stress}} = \text{strain} \tag{4.15}$$

But this relationship is linear for only small strain values. The capacitance variation ΔC can be measured using a non-resistive Wheatstone bridge of a kind similar to those used in most laboratories to measure inductances, as shown in Figure 4.16 (note the alternating current supply), thus allowing a measurement of the mechanical stress, as required.

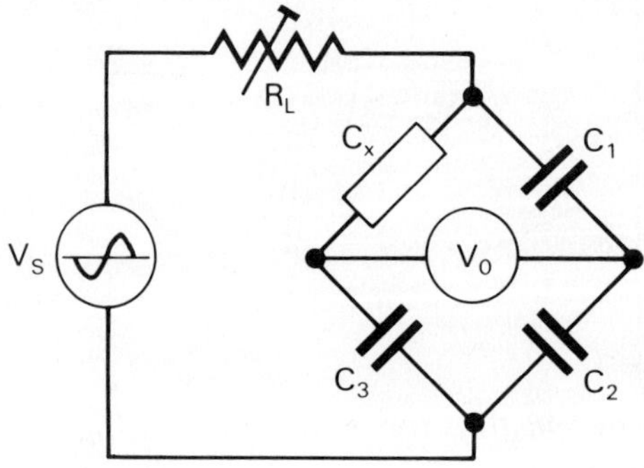

Figure 4.16 Simplified circuit diagram of a capacitor Wheatstone bridge

Other kinds of non-resistive force transducer use fibre-optic cables which detect the presence of the external physical disturbance by a change in the light transmission characteristics. Further details on this technique are reported in Chapter 7. These and other special-purpose force transducers encountered in industry suffer from limited sensitivity when compared with strain gauges and, at present, are seldom used in robotic applications.

4.5 Load cells

In practical measurements these transducers need to be mounted on a support structure which transforms the mechanical stress in to the required measurable strain. These structures are called *load cells* whose shape and general design depend on the application. An example of a load cell used in conjunction with strain gauges to measure the bending and torsion stresses acting on a robot wrist is shown in Figure 4.17(a) and (b). This load cell and

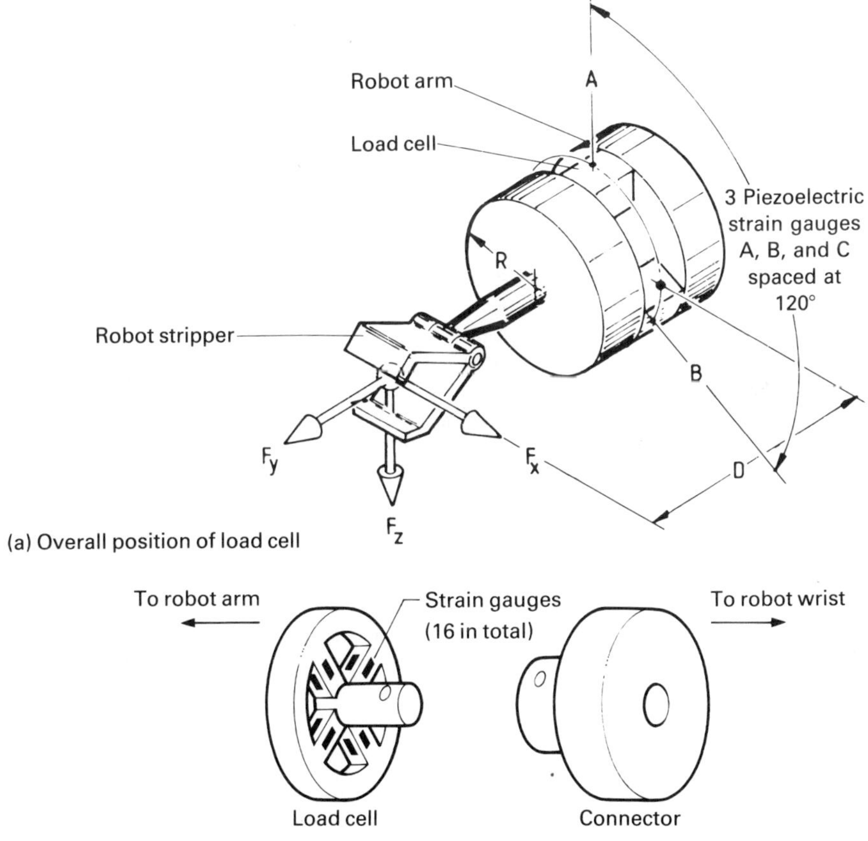

Figure 4.17 Example of a load cell using strain gauges to measure wrist forces and torques (after McCloy and Harris, courtesy of Open University press, 1986)

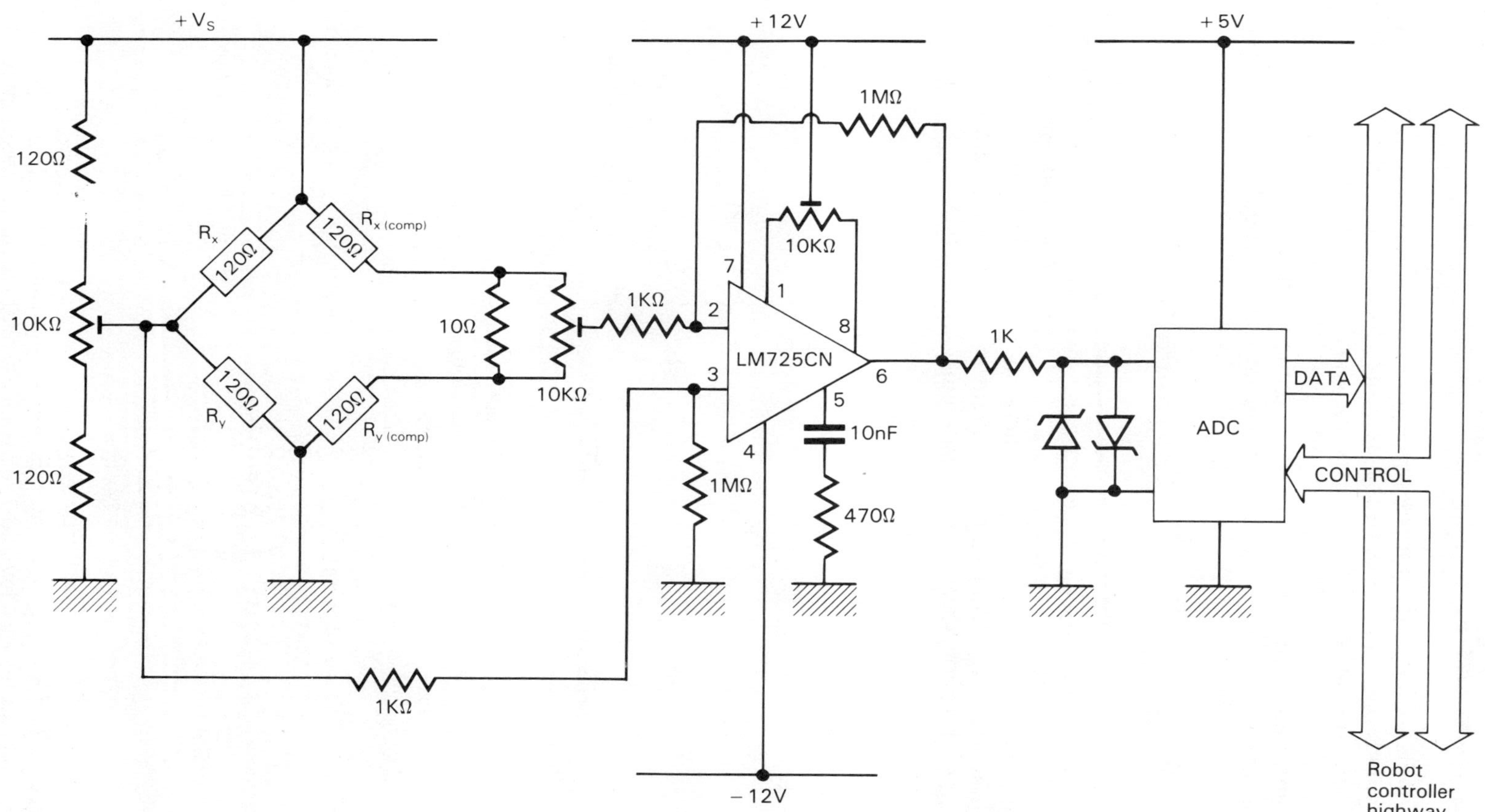

Figure 4.18 Typical low-cost computer interface circuit for strain gauge bridge

its connector are machined from a single block for stability and are available from several sources, such as the Lords Corporation in the USA and Meta Machines in the UK.

4.6 Interfacing of force transducers

The output from the Wheatstone bridge is typically of the order of microvolts and is, of course, an analogue signal. To interface it to a robot computer controller it needs A/D conversion which, in turn, requires signal amplification since the typical input range for an A/D converter i.c. is 0–2 V. The front end of the interface circuit therefore needs to provide voltage amplification of the order of 60–100 dB with very low drift and a high input impedance.

Dedicated 'strain gauge amplifier' integrated circuit, such as the SGA100, are available nowadays to meet these specifications and offer an effective solution to the problem; however they are expensive and can become unstable and so require a little experience (or strict adherence to the data sheets) if a reliable result is to be obtained.

A cheaper alternative can be implemented using low-drift, general-purpose, instrumentation amplifiers such as the LM725. An example of a strain gauge interface circuit based on such a device and providing 60 dB of front-end amplification is shown in Figure 4.18.

4.7 Conclusions

Strain gauges are the most widely used force transducer, in both the metal foil (resistive) and semiconductor (piczoresistive) types, the latter having a higher sensitivity but lower dynamic and temperature ranges. The Wheatstone bridge is the most effective technique for measuring the transducer's resistance change under the effect of the mechanical strain and therefore the force involved.

4.8 Revision questions

(a) Show the physical configuration of a strain gauge used to measure the twisting force on the structure shown in Figure 4.a.

(b) Figure 4.b shows two strain gauges used to measure twisting force. Show their electrical connections within a bridge circuit such as to reduce the effects of bending moments. (You can ignore temperature variations and axial movement.)

(c) The four gauges in Figure 4.12 are used to measure traction force; the electrical connections are as shown in Figure 4.13. The bridge operates

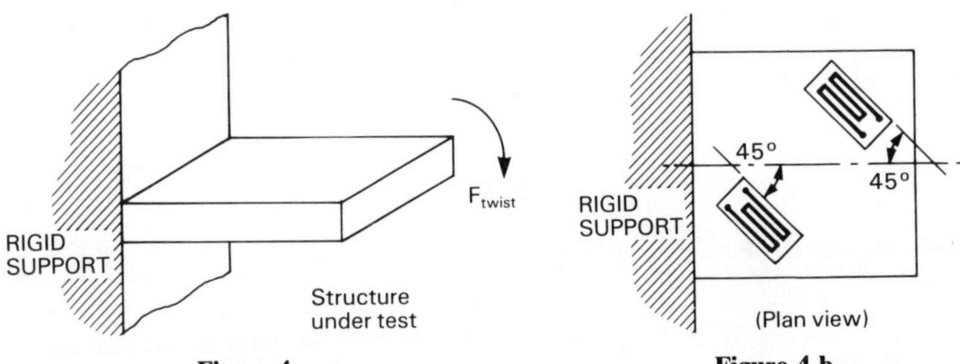

Figure 4.a **Figure 4.b**

from a 10 V supply and is initially balanced. Calculate the output voltage change ΔV_0 under a stress of 0.4 MN/m using Nichrome gauges mounted on a steel structure (assume a Poisson ratio of 0.3).

4.9 Further reading material

1. *Transducers, Sensors and Detectors,* R. G. Seippel, Reston Publishing Co. Inc., 1983.
2. *Transducers and Interfacing,* B. R. Bannister and D. G. Whitehead, Van Nostrand Reinhold (UK) Co. Ltd., 1986.
3. *Robotics—An Introduction,* D. McCloy and M. Harris, Open University Press, 1986.
4. *Industrial Robots: Computer interfacing and control,* W. E. Snyder, Prentice-Hall Inc., 1985.
5. *Industrial Robots: Design, operation and application,* T. W. Polet, Glentop Publishers Ltd., 1985.

Chapter 5

Velocity transducers

5.1 Overview

Velocity transducers are based on transforming the physical velocity (or speed) of a rotating shaft into an electrical signal. The shaft velocity v is however the derivative function of its angular position α:

$$v = \frac{d\alpha}{dt} \tag{5.1}$$

It follows therefore that it is possible to measure velocity by using a position transducer and calculating the rate of change of its output variable. This leads to a classification of velocity transducers into two distinct types as shown in Table 5.1.

The latter category, being derived from position transducers, is also based on the use of encoder disks to produce either optical, capacitive or magnetic (Hall effect) incremental transducers. The use of the potentiometer as a velocity transducer would require interfacing to a differentiator circuit to provide the position derivative; but these circuits are difficult to stabilize and their use is not recommended. The angular resolution of capacitive and magnetic encoders is poor compared with optical ones, so they do not find wide application in the field of robotics.

The optical incremental encoder and, arguably, the tachogenerator therefore remain the most popular velocity transducers in the field of machine control and robotics.

Table 5.1 Velocity transducers.

Direct	Transducers which measure the shaft velocity directly. These are analogue devices whose variable output is proportional to the shaft rotational velocity. Examples are d.c. and a.c. tachogenerators.
Derived	Transducers which are used mainly for measuring shaft position but which, by external electronic processing, can also provide a shaft velocity measurement. These are digital devices whose output is a pulse train proportional to the shaft rotation, which is then processed electronically using digital counters, to produce a signal proportional to the shaft position and/or angular velocity.

5.2 Tachogenerator

The tachogenerator is a direct-type velocity transducer. It is essentially the opposite of an electric motor since its input is a rotating shaft and its output is a voltage V_0 proportional to the input angular velocity ω_s:

$$V_o = K_t \cdot \omega_s \tag{5.1}$$

The constant of proportionality K_t is generally linear within a large dynamic range but does vary in relation to: load current, maximum input velocity and temperature. These variations are usually amply documented in the manufacturer data sheets and do not present a problem. Figure 5.1 shows a typical relationship between V_o and ω_s for different load resistors.

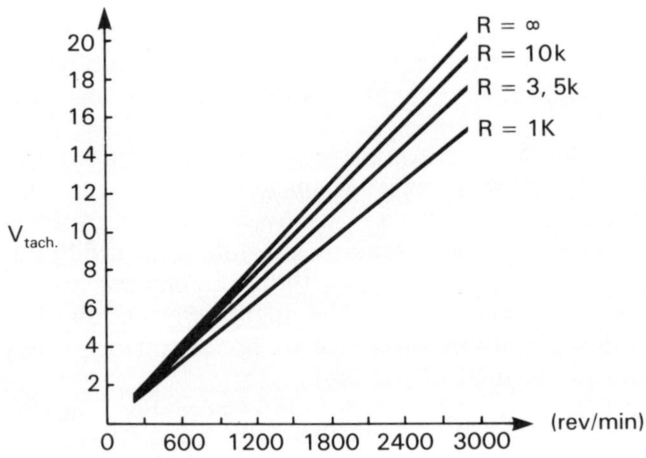

Figure 5.1 Typical transfer characteristic of tachogenerator

The principle of operation of a tachogenerator, being the dual of an electric motor, is also based on Faraday's law of electromagnetic induction which states, as shown by eqn (5.2), that: the magnitude of a voltage induced in a single loop conductor is proportional to the rate of change of lines of forces passing through (or linked with) that loop.

$$V_o = \frac{B \cdot l \cdot w}{10^{+8}} \tag{5.2}$$

where B is the magnetic flux density, 'l' the length of the active portion of the conductor linking the flux, w is the relative velocity between the conductor and the magnetic field and 10^{+8} is the number of lines a single loop must link per second in order to induce a voltage of 1 V.

In the tachogenerator case the magnetic flux density B and the loop length l are constants, so the output voltage V_o is proportional to the shaft angular velocity W_s as required and a shown previously in eqn (5.1).

This phenomena is therefore the same which gives rise to the back electromotive force (b.e.m.f.) in electric motors. The equivalent circuit of a tachogenerator is also the dual of an electric motor and is shown in Figure 5.2:

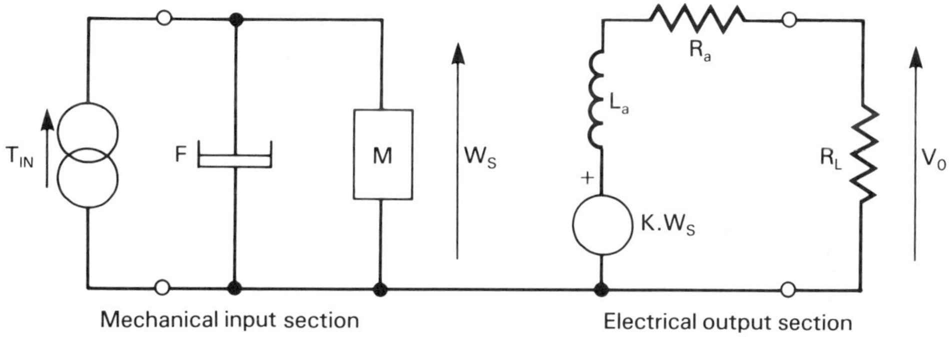

Figure 5.2 Tachogenerator equivalent circuit

where T_{in} is the torque input whereas F_a, M_a, L_a and R_a are all armature parameters, respectively friction, mass, winding self inductance and winding resistance. R_L represents the electric load connected to the tachometer. Ignoring any transients, and therefore the effects of all reactive components, we can show that the output voltage is given by eqn (5.3):

$$V_o = (K_a\omega_s - R_a i_L) = \frac{R_L \cdot K_a \cdot W_s}{(R_L + R_a)} \tag{5.3}$$

which, once again, can be simplified back to eqn (5.1) by lumping K_a and the resistances in the total constant K_t. The output voltage V_o can be in direct current form in the case of the dynamo (a d.c. tachogenerator) or alternating current form in the case of the alternator (an a.c. tachogenerator).

Compared to other velocity transducers the main advantage of a tachogenerator is that it provides an output without the need for a power supply. Its main drawbacks, on the other hand, are that it has an analogue output (which requires A/D conversion when interfaced to a computer); the output can suffer from large fluctuations, and it has a shorter operational life.

Both of the latter two drawbacks are caused by the presence of the commutator in the d.c. type tachogenerator and can therefore be reduced if an a.c. type tachogenerator is used since this one employs slip rings to extract the power from the armature windings. The a.c. output would however require rectification and smoothing before interfacing to the A/D or the use of a peak detector.

5.3 Optical incremental encoder systems

As shown previously in Section 2.3 this transducer is used mostly for position feedback and its direct output is a pulse train whose frequency is proportional to the shaft angular speed. Counting hardware needs therefore to be used in order to extract the shaft angular position or, as in this case, its angular velocity. This principle is illustrated in Figure 5.3:

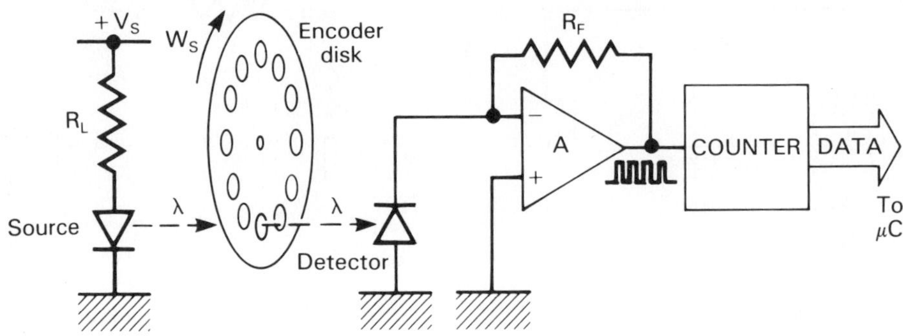

Figure 5.3 Velocity transducer system based on an optical incremental encoder

The light source is usually an Infra Red Light Emitting Diode (IR LED) whereas the light detector is a spectrally matched photodiode (IR PD); for further information on both these devices please refer to Chapter 3 on optical transducers.

The modulation of the light from the source to the detector can be achieved in different ways, as applies also to position transducers; the perforated disk shown in Figure 5.3 can in fact be substituted by a transparent (i.e. glass or clear plastic) disk with a pattern of light and dark stripes printed on it. These light stripes can, in turn, be transparent or reflective which, in the latter case, permits the use of single packaged LED-PD pairs. A comparison of these two alternatives is shown in Figure 5.4:

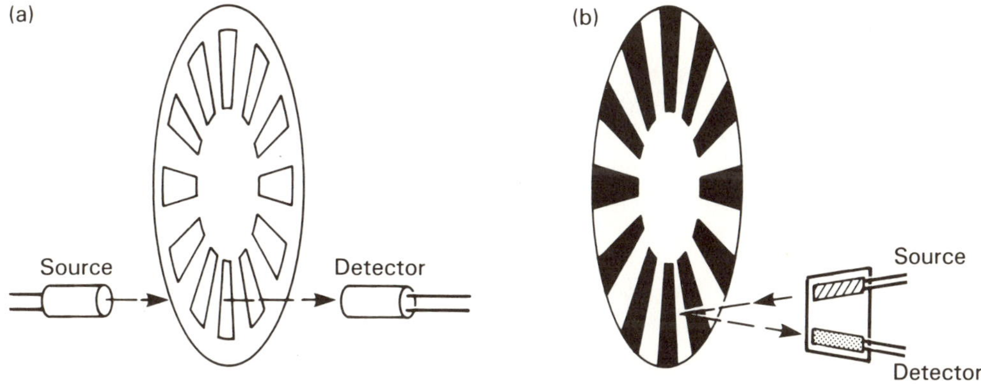

Figure 5.4 Comparison of transmissive and reflective incremental encoder disks

The advantage of using optical reflection rather than optical transmission as the light modulation principle is twofold; a more compact transducer due to the single LED-PD package and a more versatile transducer due to the opportunity to discard the encoder disk and coat (or etch) the required single pattern of alternate light and dark stripes directly on the surface to be monitored, such as the shaft of a gear box shown in Figure 5.5. In most cases, in fact, the encoder disk used in velocity transducers, unlike the position transducer's one (see Section 2.3.1), does not need a second phase-offset light modulating pattern and a reference slot so as to determine the direction of rotation and number of revolutions. We can, in fact, measure the speed, a scalar quantity, with a single disk pattern and convert it to velocity, a vector quantity, by adding the direction of rotation. This can be

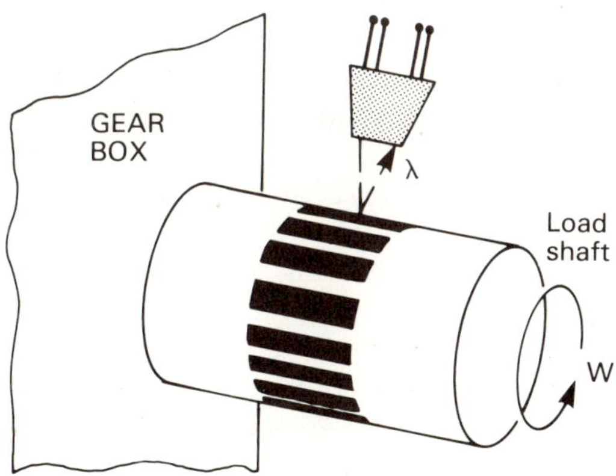

Figure 5.5 Example of direct load shaft speed monitoring using reflective optical transducer

done within the computer either as a measurement from the position transducer (that is the velocity transducer supplies the vector modulus and the position transducer the direction) or as an estimated data (that is the computer predicts the direction of rotation from the coordinate it generated for the motor operation). This latter technique, however, is not suitable for robot systems where load position oscillation is likely to occur, as is the case when driving large inertial loads.

5.4 Interfacing of velocity transducers

The *tachogenerator interfacing* is quite straightforward since it requires only A/D conversion, though it may, in some applications, require low-pass filtering to remove commutator spikes and/or output voltage attenuation to match it to the A/D working range. An example of interfacing a tachogenerator to a digital system is shown in Figure 5.6.

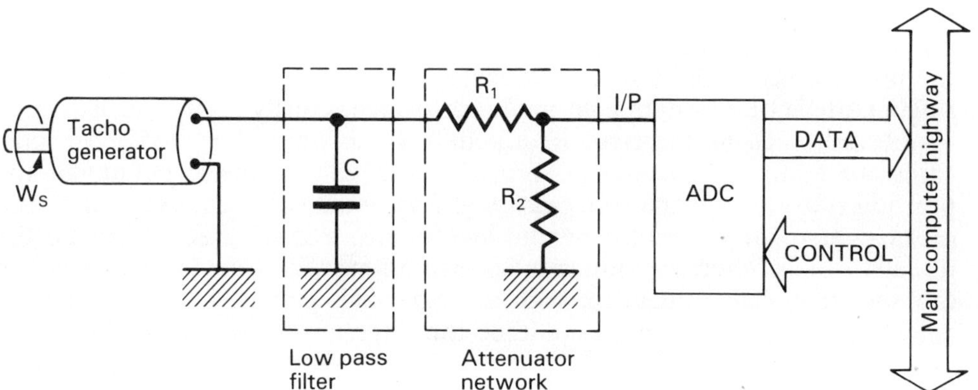

Figure 5.6 Generalized computer interface circuit of a tachogenerator

Since the transducer output is analogue (i.e. continuous) the velocity measurement resolution depends only on the number of bits used by the A/D converter; an 8-bit converter will therefore provide just under 0.4% measurement resolution, that is $(1/256) \times 100 = 0.39\%$.

Optical incremental encoder interfacing, on the other hand, requires a voltage amplifier as well as a counting circuit in order to interface it to a computer. The optical encoder output is a low-voltage pulse train whose frequency is proportional to the shaft angular speed. To measure this speed the interfacing hardware needs to count the number of shaft revolutions per unit time. This can be achieved in two ways:

- By counting the number of lines detected by the photodiode in a unit time, a technique suitable for measuring medium to high speeds.
- By measuring the time taken by the photodiode to detect a single line.

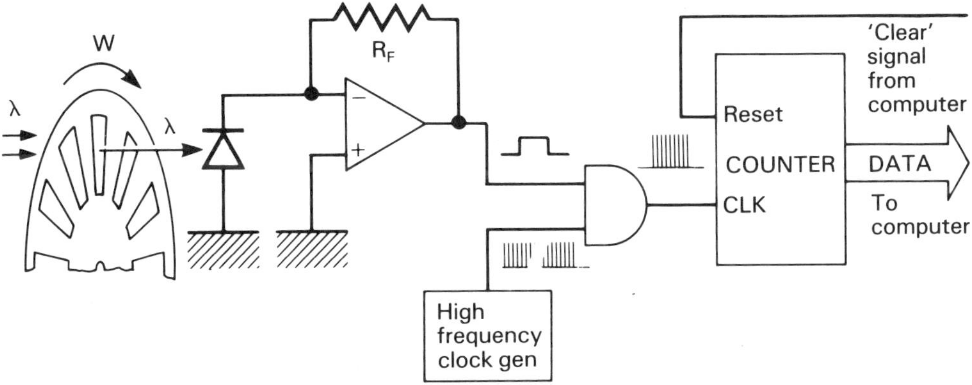

Figure 5.7 Simplified circuit diagram of low speed measurement using single-line pulse technique

The latter is a useful technique when measuring low speeds with high resolution and is based on gating a high-frequency clock with the photodiode output pulse corresponding to a single line transition (Figure 5.7).

The former technique is, however, more useful because in most cases robot shaft actuators rotate at medium to high speeds (that is in excess of 1 rev/s) and will therefore be described in more detail.

To measure speed by the former technique, one pulse per revolution (i.e. a single line or slot on the encoder disk) would be sufficient. But digital counters have an inherent count resolution of ± one clock pulse so, in order to allow the accurate measurement of low angular velocities, more than one pulse per revolution needs to be counted per unit time and therefore several dark lines (or slots) are required on the encoder disk. The velocity measurement time can thus be reduced at the expense of having more lines on the disk and vice versa.

The three main components of the processing hardware are therefore an amplifier, a digital counter and a clock circuit. A real-time clock is not strictly necessary since the measurement can be calibrated on any time interval to give the output in revolutions per minute as required. However, with the advent of very stable and highly accurate quartz crystal oscillators a real-time clock is quite easily built. A circuit is illustrated in Figure 5.8

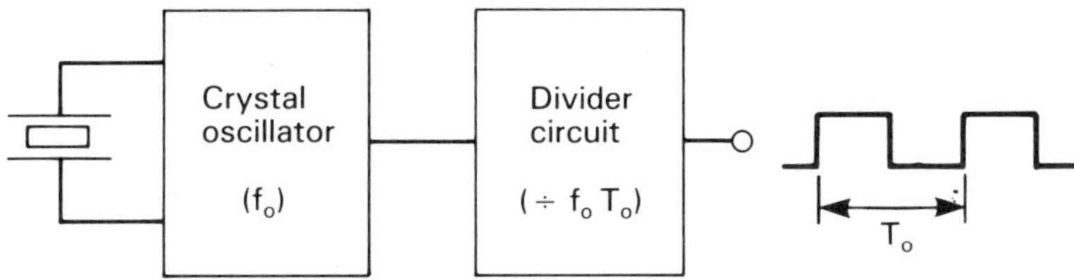

Figure 5.8 Typical block diagram of real-time clock based on a quartz oscillator

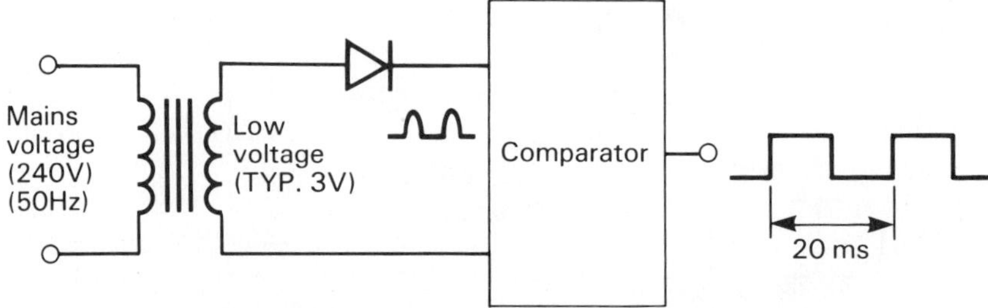

Figure 5.9 Mains synchronized real-time clock

where the output waveform period T can be made a multiple or a subdivision of a second by choosing a suitable crystal frequency and divider factor. Examples of such circuits are found in modern quartz watches.

A much less complex and cheaper alternative to a real-time clock is provided by the mains supply frequency. This is maintained stably in Europe at 50 Hz and in USA at 60 Hz. Synchronization to the mains frequency for timing purposes is therefore very convenient and easily attained, as shown in Figure 5.9.

Using the mains frequency to generate the required time intervals, as shown previously in Section 5.3.1, we can therefore design a suitable circuit for interfacing an optical incremental encoder to a robot control computer. An example of such a circuit is shown in Figure 5.10.

The output from the incremental encoder is a sequence of pulses which are counted to provide the velocity measurement, whose resolution therefore depends on the minimum number of pulses counted in the given measurement interval; that is on the minimum number of lines counted at

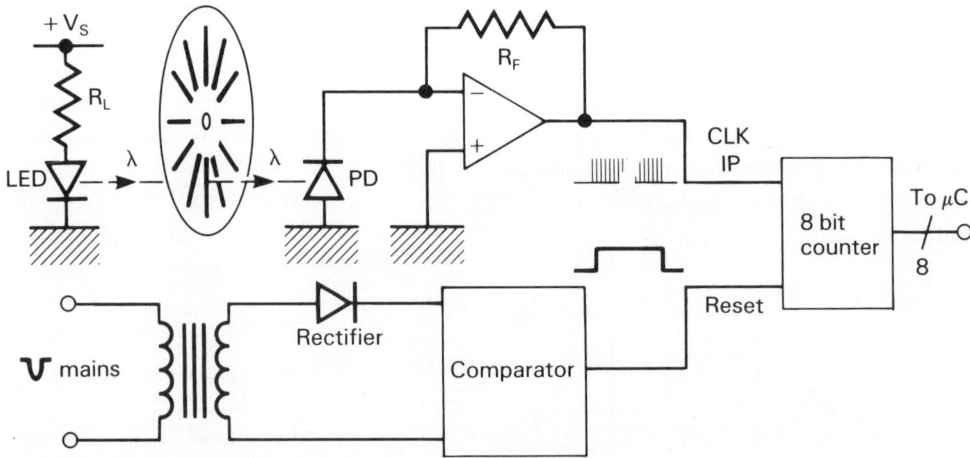

Figure 5.10 Typical velocity transducer system based on an incremental encoder disk

the slowest shaft speed. The measurement time interval itself depends on the computer handshaking, that is on how often the computer needs a velocity measurement. This time interval can be exceeded if the measurand has a longer mechanical time constant since the shaft speed would not change significantly between computer updates due to the shaft mechanical inertia.

If the computer requires a velocity measurement every 100 μs for instance, and the measurand mechanical time constant is 1 second (e.g. a shaft driving a large inertial load), then the speed will not change significantly in 100 μs and the measurement interval can be much longer, say 100 ms in this example, and the computer will simply be given 'old' measurement values during the 100 μs access times until the next measurement is carried out on the 100 ms mark. Given a measurement interval time, however, the measurement resolution depends on the number of pulses counted in that interval, hence:

$$N = L f_{\min} t_{min} \tag{5.3}$$

where

N = required number of pulses = 100/required measurement resolution
L = number of lines on the encoder disk
$f_{\min}$ = minimum frequency = minimum shaft speed in rev/s
$t_{\min}$ = measurement interval

For instance, for a 5% resolution (i.e. a maximum of 1 'wrong' unit count in 20) we need a minimum of 20 pulses counted during the measurement interval. Using a disk with 60 lines on it we can calculate the minimum shaft speed for which we can guarantee the required resolution, that is 0.3 rev/s which is 18 rev/min:

$$\text{Minimum shaft speed} = \frac{1}{f_{\min}} = \frac{L t_{\min}}{N} = \frac{60 \times 100 \times 10^{-3}}{20} = 0.3\ \text{rev/s}$$

5.5 Conclusions

The d.c. tachogenerator and the optical incremental encoder are the most popular types of velocity transducer in the field of machine control and robotics.

The tachogenerator can be interfaced easily to a computer via an A/D converter, requires no power supply nor, typically, any voltage amplification and its measurement resolution depends only on the A/D number of bits.

The optical incremental encoder requires output amplification and a digital counter for interfacing to a computer and its measurement resolution depends on the number of lines on the encoder disk as well as the measurement interval. However, it has the inherent advantage of providing both position and velocity measurement from a single transducer, though requiring the use of two separate interface circuits to achieve it, and is therefore a more versatile transducer.

5.6 Revision questions

(a) Draw the interface circuit between a computer and the tachogenerator whose characteristics are shown in Figure 5.1 using an A/D with an input range of 0–2 V. State suitable component values for a speed range of 0–1900 rev/min assuming the input of the A/D to be negligible and that no low-pass filtering is required.

(b) A photodiode circuit with a risetime of 10 μs is sued to detect the light modulation through an optical encoder disk with 1000 lines on it. Calculate the maximum shaft speed possible to maintain measurement reliability. Assume that the minimum output swing required for reliable operation of the circuit is 20% of full output.

(c) A robot control system requires a shaft velocity measurement in the range 60–3600 rev/min with a resolution of 10%. The handshaking protocol allows communication to the computer only every 10 ms. Calculate the minimum number of lines required on the encoder disk if an optical incremental encoder is to be used, and draw the block diagram of a suitable system.

5.7 Further reading material

1. *Transducers, Sensors and Detectors,* R. G. Seippel, Reston Publishing Co. Inc., 1983.
2. *Transducers and Interfacing,* B. R. Bannister and D. G. Whitehead, Van Nostrand Reinhold (UK) Co. Ltd., 1986.
3. *Robotics—An Introduction,* D. McCloy and M. Harris, Open University Press, 1986.
4. *Industrial Robots: Computer interfacing and control,* W. E. Snyder, Prentice-Hall Inc., 1985.
5. *Industrial Robots: Design, operation and application,* T. W. Polet, Glentop Publishers Ltd., 1985.

PART II
SENSORS

Chapter 6

Robot vision sensors

6.1 Overview

The last few years have seen the increasing use of robots in the industrial world and the emergence of robotics as a subject area with its own identity. The development of robot technology has been identified as following three major conceptual stages which, in turn, have helped to identify the robots belonging to these stages as respectively, robots of the first, second and third generations (Pera, 1981).

First generation robots: These are robots without any external (i.e. exteroceptive) sensors or transducers. They do not therefore have the means (i.e. the sensors) nor, usually, the computing power to interact with their environment. These robots control the end-effector by calculating its location using the data supplied by the internal (i.e. proprioceptive) position transducers present within each robot joint. For further information on the position transducers please refer back to Chapter 2.

At present the majority of the commercially used robots belong to this category.

Second generation robots: These robots have some exteroceptive sensors and a limited amount of computer power which allows them to process the environmental data and respond to it in a limited way. At present only a small percentage of the existing industrial robots can be classified under this category and their development is still largely at an experimental stage.

Third generation robots: Robots with sensors and extensive computing power which allows them to interact fully with the environment, that is: to make decisions, plan and execute the tasks that circumstances require. These robots are not yet in existence.

Second and third generation robots therefore require sensors that can provide the necessary environmental feedback and help to increase the robots' accuracy and/or their flexibility. The development of these sensors can follow two main paths:

(i) a long term strategy aimed at the development of a general-purpose, flexible sensor
(ii) a shorter term strategy aimed at providing specific, albeit inflexible, solutions to current industrial automation problems.

The former option is more suitable for the development of the future Flexible Manufacturing Systems but does require a larger investment, not only in terms of capital but also, quite significantly, in terms of manpower and planning. This type of research therefore tends to be limited to academic institutions and large corporate companies.

The latter option, by contrast, reflects the needs of western industries to be both competitive and efficient, achievements which are measured on relatively short time scales.

It comes as no surprise, therefore, to find that in the field of Robotic research the majority of USA and, to a lesser extent, European research institutions, are following a plan of 'task driven' research based on a close collaboration between the academic institutions and individual 'pools' of industrial sponsors. This arrangement, first pioneered by Rosen *et al.* at the Stanford Research Institute in the early 70s, is believed to be one of the underlying causes for the present USA success in this field.

The choice of which type of sensor is incorporated in the robot control structure depends, on course, on the application. It is generally accepted, however, that vision is the most powerful and yet flexible type of environmental feedback available, which has led to considerable research and development in to this field. Indeed Robot Vision and Sensory control is now an established conference topic in its own right. Table 6.1 shows the main areas of interest in the field of robotic vision.

6.2 Illumination considerations

Object illumination is an important element of the visual data acquisition. Its purpose is to provide adequate distinction between the background and the target object (e.g. a car component on a conveyer belt) and help to bring out the relevant features of the object.

In the case of passive vision systems, such as those using cameras, object illumination can be very critical and ordinary ambient lighting, whether artificial or natural, is quite unsuitable since it may be either too low or produce misleading shadows and/or highlights.

Table 6.1 Robot vision systems

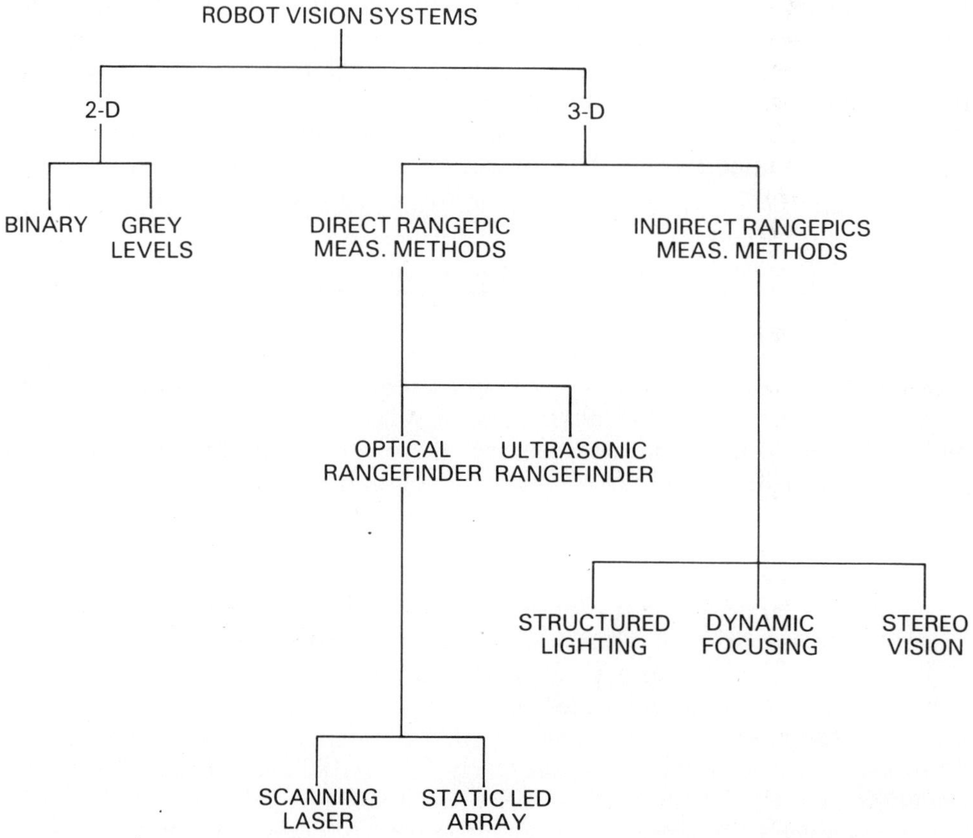

- ROBOT VISION SYSTEMS
  - 2-D
    - BINARY
    - GREY LEVELS
  - 3-D
    - DIRECT RANGEPIC MEAS. METHODS
      - OPTICAL RANGEFINDER
        - SCANNING LASER
        - STATIC LED ARRAY
      - ULTRASONIC RANGEFINDER
    - INDIRECT RANGEPICS MEAS. METHODS
      - STRUCTURED LIGHTING
      - DYNAMIC FOCUSING
      - STEREO VISION

Active vision sensors, such as direct 3-D rangefinders, produce their own object 'illumination', whether optical or ultrasonic, and are therefore not dependent on specially designed, and sometimes critically adjusted, artificial lighting systems.

In general, however, for passive vision sensors, two main techniques of illumination are employed:

- reflected illumination
- through illumination

Both these techniques present advantages and disadvantages and are used according to specific requirements in each case.

Reflected illumination yields details of the surface features of the object. However, to extract the required information from the picture, careful attention to lighting or powerful data processing software must be used to cope with the multigrey level image. Reflective illumination also includes diffused lighting to avoid shadow formation, and stroboscopic lighting to 'freeze' moving objects.

Through illumination allows easy distinction of the dark object on a light background (and vice versa). However, only the contour and 'through-going' features of the object provide data for the subsequent processing and analysis. Furthermore through illumination produces only binary images (i.e. a pattern of black and white pixels) while in the case of reflected illumination the image can be digitized into any required number of grey levels thus yielding more information.

A comparison of the two illumination techniques is summarized in Table 6.2 while Figure 6.1 provides a pictorial description of the different modes of illumination (from PERA Report No. 366).

Table 6.2 Comparison of two main illumination techniques (Courtesy of PERA)

	Reflected illumination technique	*Through illumination technique*
Principal advantages	Possibility of surface features for image analysis; grey scale gradations are possible; additional information about object's shape is available; use of code markings possible	Easy separation of objects from background; minimum amount of data to be handled
Principal disadvantages	Extracting the relevant picture detail may be difficult and slow; shadows and highlights have to be considered; extraneous light sources may influence system functioning	Only contour and 'through' features of object are accessible to image analysis; transparency of object support, viewing and illuminating elements can be impaired by dirt, etc.

6.3 Vision sensor generalities

Vision sensors are so called because they possess, in their make-up and functioning, a certain analogy with the human eye and vision. The analogy is somewhat easier to see in the case of vacuum and solid-state cameras, because they also possess the 'equivalent' of the human retina in the form of a photosensitive array. In the case of some active 3-D vision sensors, such as scanning laser rangefinders, the data is acquired by mechanical scanning using a single optical transducer.

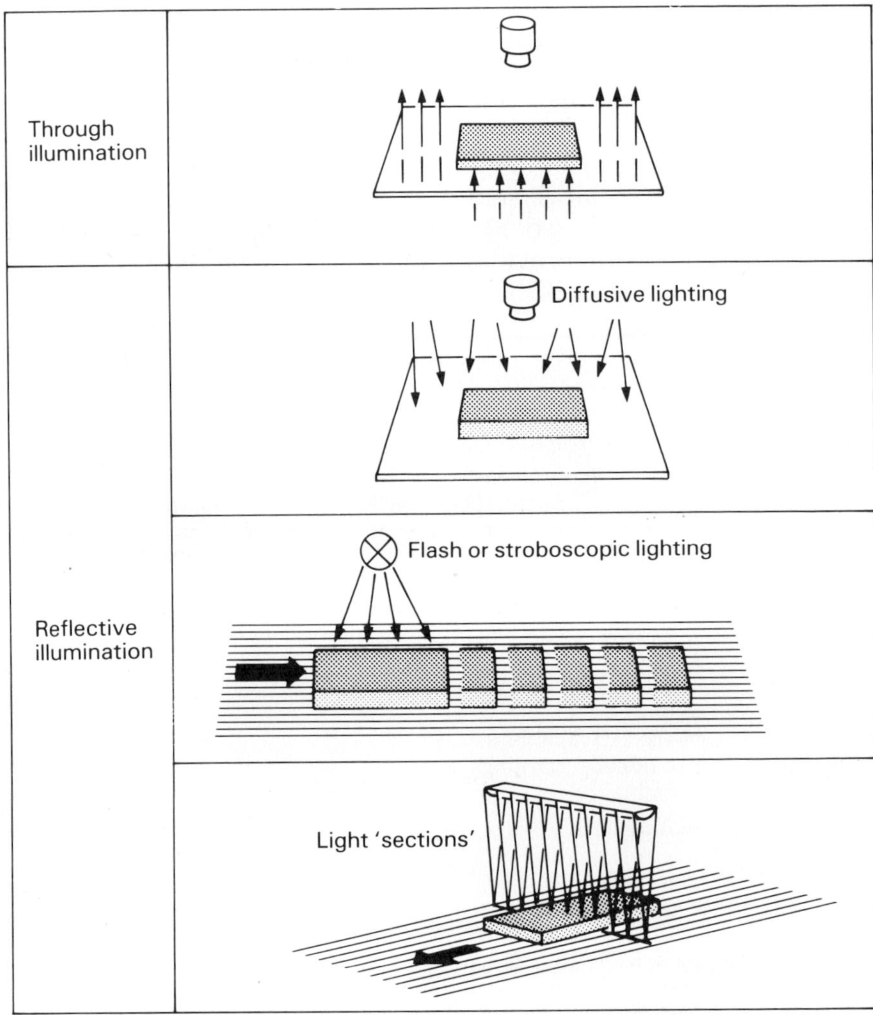

Figure 6.1 Different modes of illumination (courtesy of PERA)

In spite of their differences, however, all vision sensors can be broken down into the same constituents. Figure 6.2 shows a typical block diagram of a vision sensor.

6.4 2-D sensors

In the USA, as in Europe, there is still considerable interest in 2-D robot vision, both in terms of an effective and practical, though limited, implementation of visual sensory feedback—there is in fact a school of

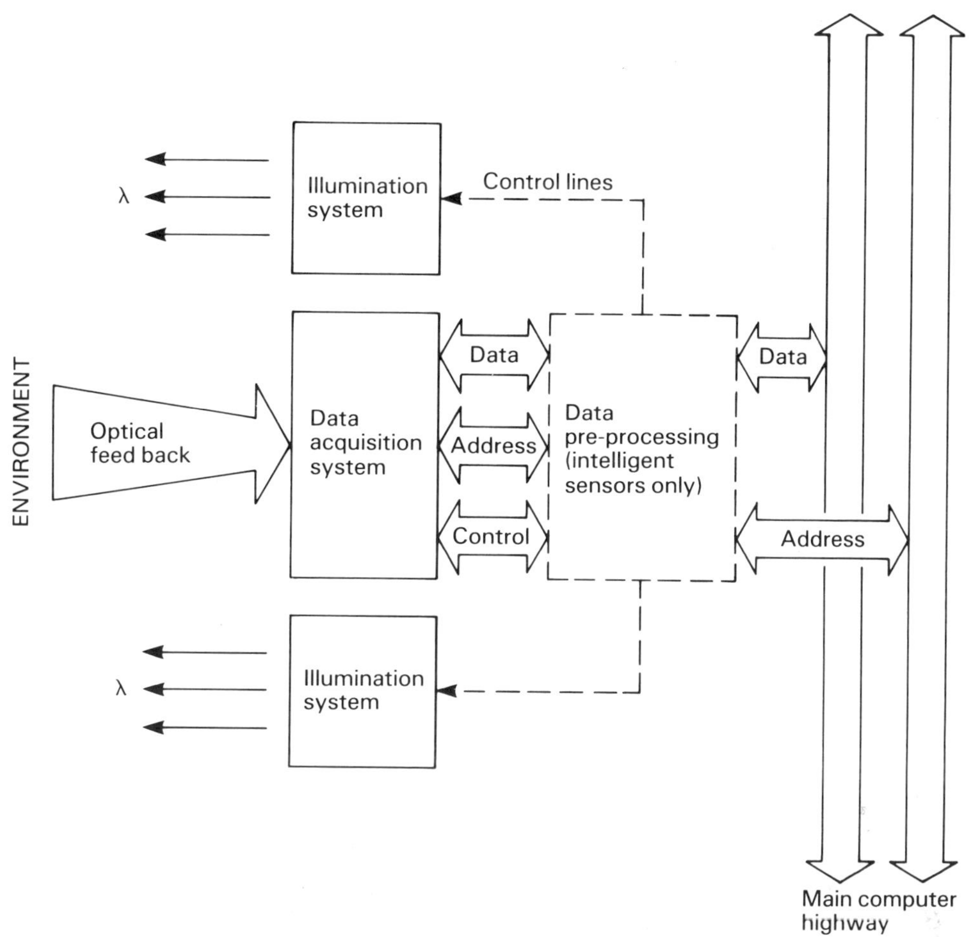

Figure 6.2 Vision sensor, generalized block diagram

thought that prefers 2-D to 3-D vision systems in view of their simpler and more trouble-free operation (Keller, 1983; Corby, 1983)—and as the input stage for 'model-based' vision processing systems that aim to derive a 3-D image from the 2-D data using a suitable model of the object in question (Meyer, 1984). The research emphasis, however, is on 3-D robot vision as a means of achieving faster, more accurate and more flexible robot operation.

2-D vision sensors are based on optical array transducers (both vacuum and solid-state types, such as camera tubes, CCD, DRAM, and photodiode arrays), dealt with in Chapter 3, and specially designed artificial lighting. Some 'intelligent' 2-D vision sensors also possess limited computer power to enable them to carry out a certain amount of picture pre-processing. This may be used both to relieve the data-processing burden on the main control computer and to interact with the illumination system (e.g. to control the lighting level) and so provide the best possible picture quality to the main computer.

Most optical array transducers (e.g. vacuum and solid state cameras) have the capability, though via different means, to provide a 'grey level' output image. The only exception to this guideline is the DRAM camera (for further details, see Section 3.5.2.4) which, being based on a memory device, produces an inherently binary output image. It should be pointed out, however, that a grey level image can be produced with a binary device at the expense of the image acquisition time by overlaying several binary images obtained with different exposure times.

The lighting conditions and/or the computing power limitations may,

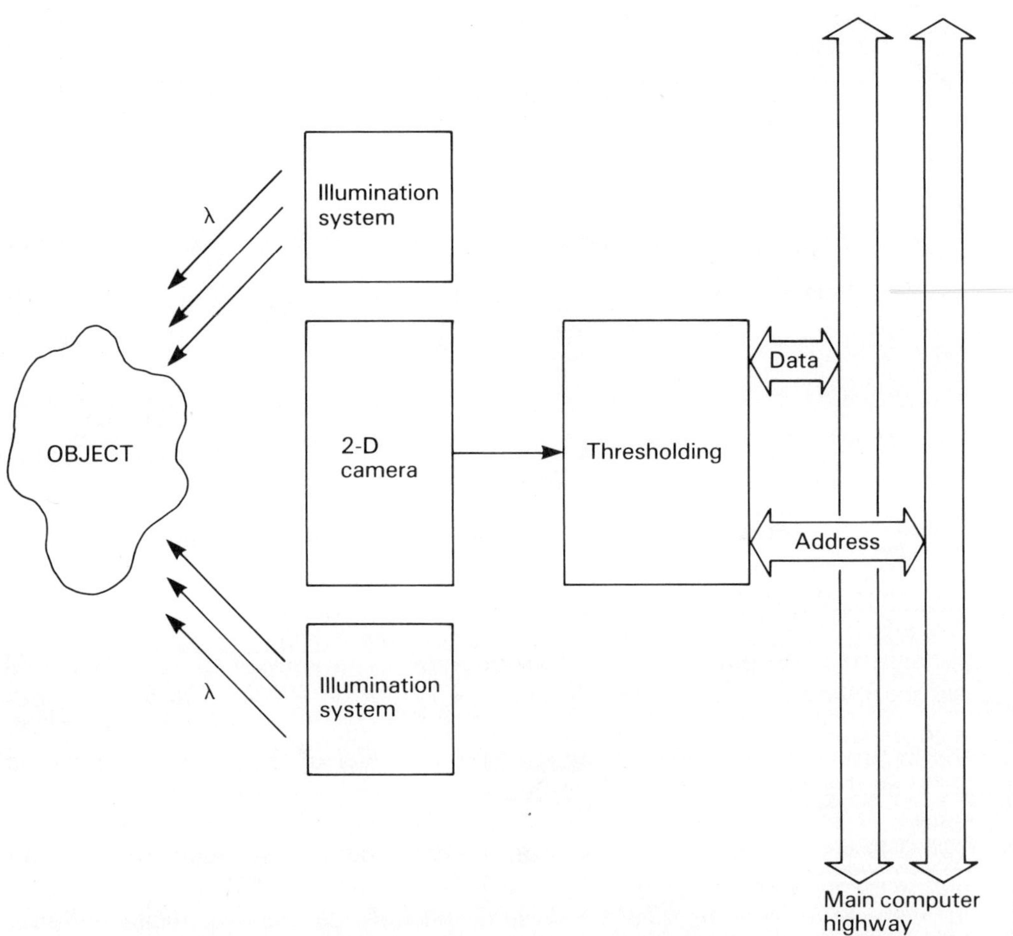

Figure 6.3 Block diagram of typical binary vision sensor

however, be such that the transducer output needs to be interpreted as binary anyway. This is the case for through illumination which, for opaque objects, helps to produce images with only two distinct grey levels. Another such example is the use of microcomputers for robot control and image processing: these devices cannot, in fact, easily handle grey level images because of their large memory requirement and therefore often need to threshold the transducer output (a 256×256 image with a 8-bit grey level resolution would in fact require 65,536 bytes of memory to store it). 2-D vision sensors can be therefore divided into two main groups:

(i) binary
(ii) grey level.

6.4.1 Binary vision sensors

This class of sensor produces images whose pixel (pixel = picture element) values are either a black or white luminosity level equivalent to a logic 0 or 1, hence the name 'binary'. Figure 6.3 shows a typical block diagram of a binary vision sensor:

The complete picture is therefore only a series of logic 1 and 0. This allows easy distinction of dark objects on light background (and vice versa) and, in view of the low visual data content, fast image data manipulation such as object perimeter and/or area calculations.

Figure 6.4 shows the typical output of a complete binary vision system:

```
LENSCAP     0.0     9.8    15.8     47.0    153.0
BOLT        0:0     1:4    55:5    167.0    143.0
WASHER      1.0     1.0    14:9     96.0     76.0
            0.0    33.0    15.6     95.0     70.0
```

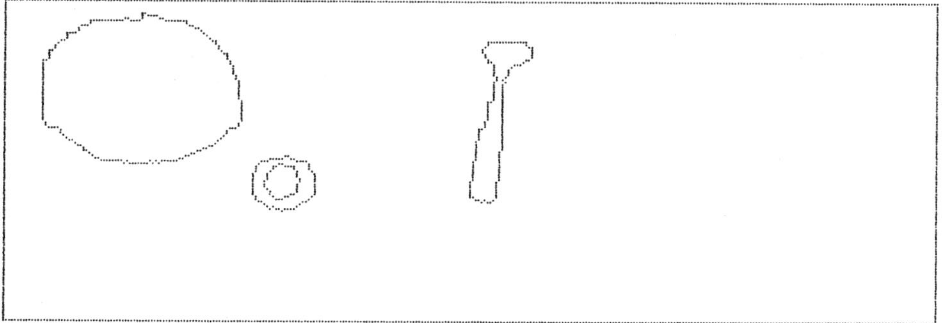

Figure 6.4 Typical example of an object recognition program based on a binary image

In the example shown the vision system has learnt the characteristics of three objects, based on the number of holes, object size and shape features. Thereafter, whenever the vision system is presented with an image containing one of these objects it will be able to 'recognize' it, that is acknowledge its presence within the field of view. This is achieved by comparing the features of the object in the new image with those stored in memory, looking for a match. In practice a certain amount of tolerance needs to be built into the matching algorithm because the sensing process is not perfect and will not always produce the same output for the same object, so that a 'perfect' match would hardly ever be possible. Too much tolerance, however, may result in the vision system 'recognizing' two similar objects as being one and the same. Further details on image degradation and noise are given in Chapter 8 on image processing.

This technique of image processing is sometimes referred to as the SRI method (or one of its derivatives) because it was first introduced by Rosen *et al.* at the Stanford Research Institute (now SRI International) and imposes severe restrictions on the lighting conditions and on the position of the object for a successful operation. The object must, in fact, show up as an isolated silhouette and its stable states must be known in advance (that is the object can be rotated within the image plane but not tilted with respect to it). Thus overlapping objects and situations with any 3-D freedom of movement are difficult to deal with using binary vision sensors.

6.4.2 Grey-level vision sensors

This class of sensor produces an output image whose pixel values are quantized into a number of discrete levels, achieved by converting the optical transducer analogue output into the appropriate number of digital levels.

An alternative to this method is necessary when using a DRAM camera as the imaging device because its output is inherently binary; a grey-level image can be produced with the help of some computing power (either within, as in the case of intelligent sensors, or without, by using the vision system main computer). This procedure is based on processing an appropriate number of binary images obtained using different exposure times, whose difference provides a measure of each pixel intensity in the object image.

Figure 6.5 shows the typical grey-level image of a bicycle chain link using eight intensity levels as provided by a 2-D grey level vision sensor based on a 256×128 cells DRAM camera. A comparison of this image with the one obtained using a binary sensor, as seen in Section 6.4.1, shows how grey-level vision sensors (in conjunction with reflected illumination) provide more details of the object surface features. This, however, increases the amount of visual data and makes its processing more difficult, slower and computationally more expensive.

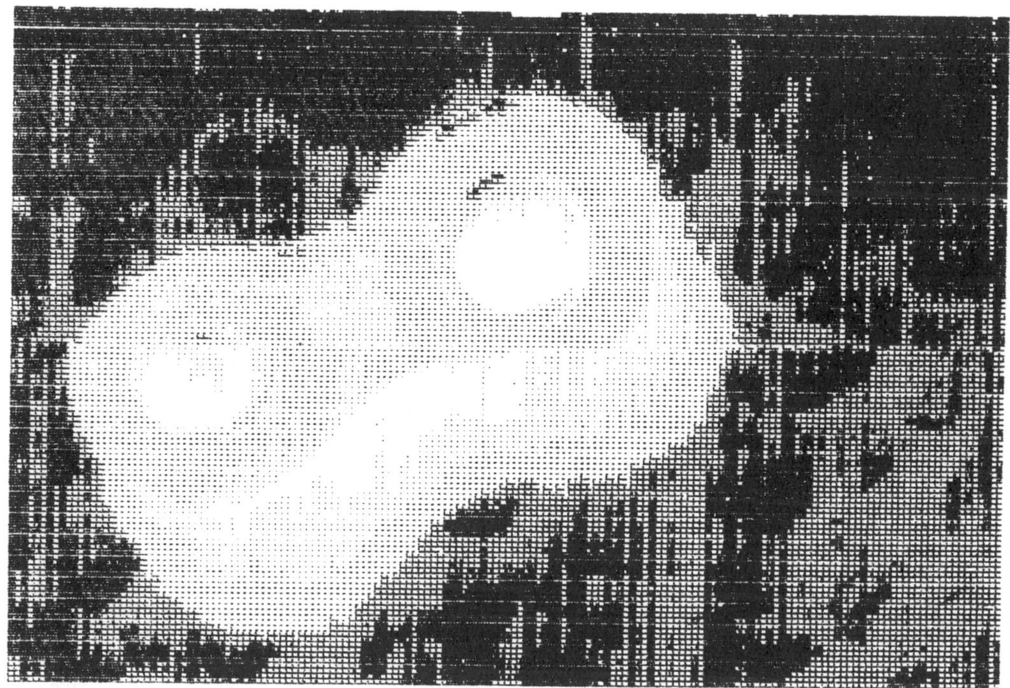

Figure 6.5 Example of grey-level image

6.5 3-D Sensors

Indirect methods of obtaining depth maps, based largely on triangulation techniques, have provided the largest input in this area. This is thought to be due in part to the existing optoelectronics technology (camera tubes, photodiode arrays, etc.) which, being inherently 2-D devices, require triangulation techniques to integrate them within 3-D vision system, and in part to the analogy with the human vision system which is also based on a triangulation technique (Marr and Poggio, 1976, 1977).

6.5.1 Stereo vision

Stereo vision, in particular, has received considerable attention. The disparity technique is based on the correlation between images of the same object taken by two different cameras under the same lighting conditions (Marr and Poggio, 1976), while the photometric technique is based on the correlation between the images taken by the same camera under two different lighting conditions (Ikeuchi and Horn, 1979).

Stereo vision sensors, like their 2-D counterparts, are also based on

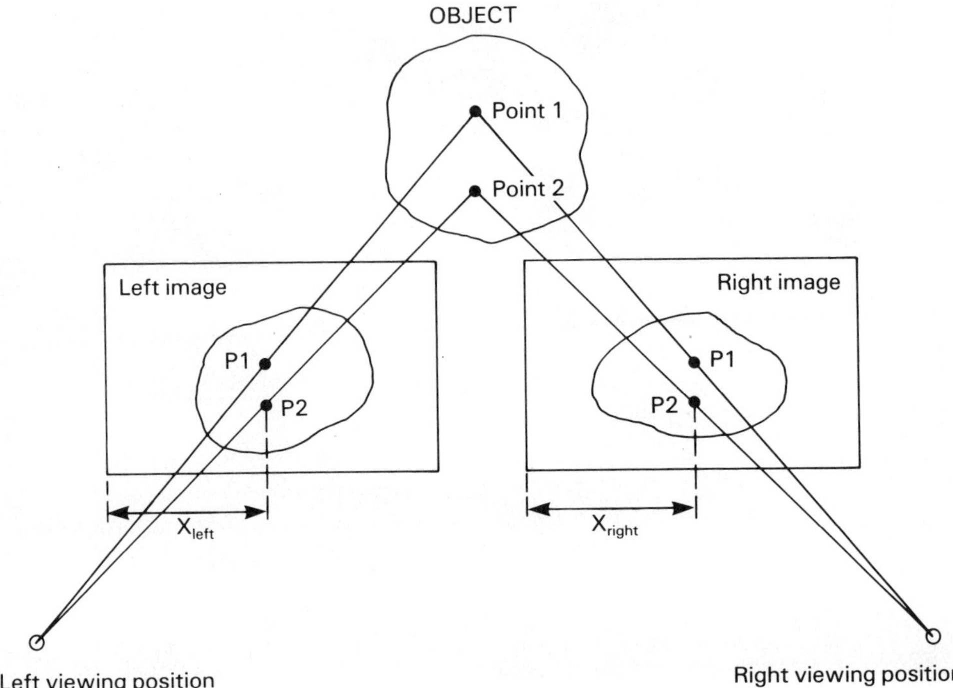

Figure 6.6 Stereo imaging diagram (after Nevatia, courtesy of Prentice-Hall, 1982)

optical array transducers, both vacuum and solid-state (such as camera tubes, CCD, DRAM and photodiode arrays). Their main function is to provide multiple position views (usually two) of the object in question. Figure 6.6 shows a diagrammatic view of how these two images are obtained.

To draw the imaging lines in Figure 6.6 (which, for simplicity's sake, were limited to two per image) one must consider that each point of the object's image corresponds to one point on the object surface (assuming properly focused optics). This means that this object point must lie along the line joining the image point and the focal point of the imaging device lens, its distance along the line being unknown. If the object is now viewed from a different angle *and the same point is visible in both views,* then it must lie at the intersection of the lines determined from the two separate views; its position (i.e. the distance from the imaging devices) can then be calculated by triangulation.

For the computer to carry out this calculation automatically, however, it needs to process the two images in three main steps:

1 Determine the point pairs in the two images, that is, determine which point in the right image corresponds to which point in the left image. This is the hardest, and therefore computationally the most expensive,

part of stereo vision. It may, in fact, be very difficult to identify the same features in both images. The image of a small area on the object surface may be different in the two images because of the different perspective and surface reflectivity due to the viewing angles. Moreover, some of the points in one image may not be visible in the other.

2 Translate the corresponding two points in the left and right images to yield the disparity measurement (i.e. the difference in the $x-y$ position of the point in the left image compared with the $x-y$ position of the corresponding point in the right image).
3 Determine the distance of the object point from the imaging devices by triangulation. (This operation requires data on the relative positions and orientations of the stereo imaging device(s) which produced the left and right images.)

The essence of stereo vision, therefore, is step 1, namely the solution to the disparity problem. To help gauge the difficulty attached to such a step one needs to note that all disparity measurements computed using local similarities (or features) may be ambiguous if two or more regions of the image have similar properties.

Consider for example the left and right images as shown in Figure 6.7

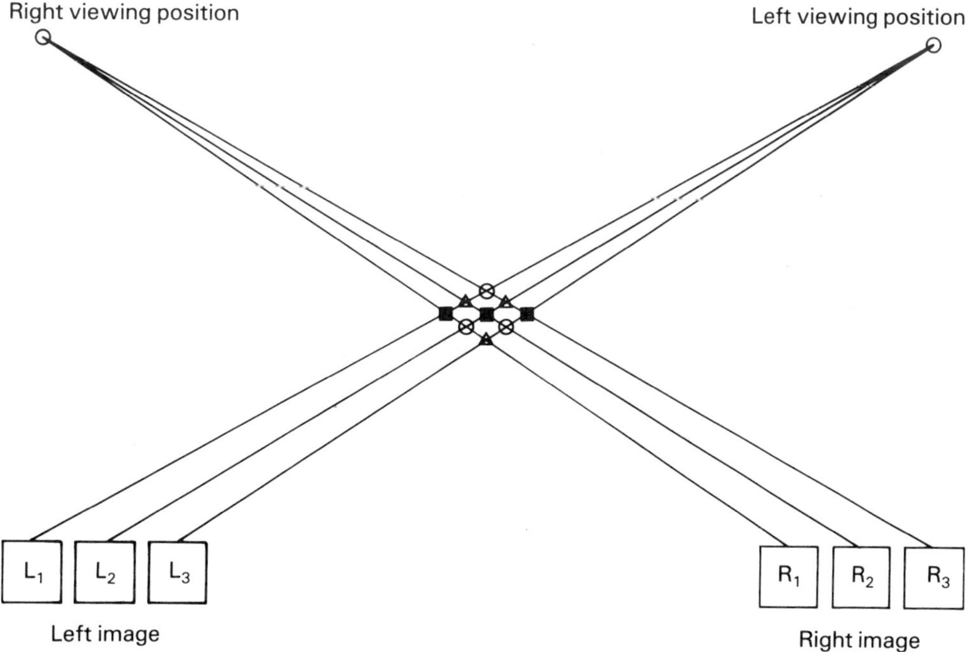

Figure 6.7 Stereo ambiguity problem (after Nevatia, courtesy of Prentice-Hall, 1982)

consisting of three dark squares each as marked. Each square in one image is similar to any of the three in the other. If we now correspond L_1 and R_1, L_2 and R_2, L_3 and R_3, the three squares will be computed to be at the same height above the background, as shown by the filled squares. If L_1 were to be matched with R_2, L_2 with R_3 and L_3 with R_1 then the computed heights would be shown by the empty triangles. Another possible interpretation is shown by the unfilled circles, thereby giving an indication of how critical the correspondence problem can become in the absence of any known and unique object features.

In spite of these difficulties and the relatively high expected price tag, robot stereo vision is a desirable goal. Stereo vision has the highest inherent 3-D image resolution (limited only by the type of camera and its optics) and flexibility (for instance it is the only method that can provide colour images relatively easily) and as such it comes closest to the aforementioned definition of a general-purpose, flexible vision sensor (see Section 6.1). This makes it a desirable goal but does require large investments and long project lead times.

The USA financial investment in stereo vision research, for instance, has already been considerable (approx. $3,000,000 to date), but the results so far have been limited mostly to laboratory prototypes. The reasons are thought to be many and varied, ranging from the aforementioned difficulty of solving the disparity problem in a sufficiently short time to the sheer complexity of a system that is essentially trying to emulate a major function of the human brain. Reccntly, however, there have been reports of successful industrial applications, such the positioning of car bodies, using stereo vision (Rooks, 1986).

There are 3 main methods of using the 2-D vision sensors to obtain multiple views as required for stereo vision:

(a) *Disparity method* 1. Use of two stationary imaging devices. This

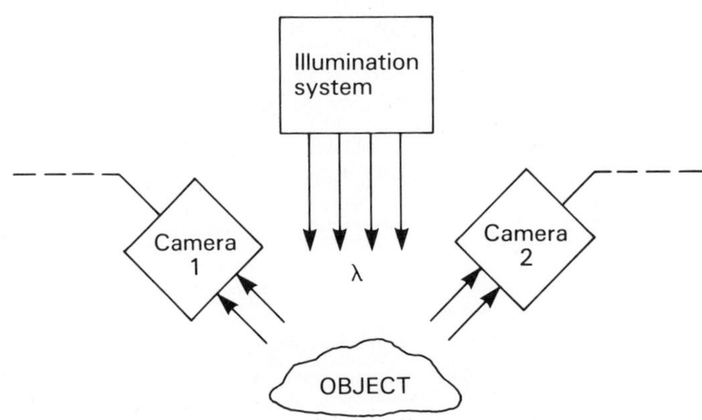

Figure 6.8 Stereo vision, disparity 1 method

could be defined as the 'classical' stereo vision method because of its analogy to the human vision system. As shown in Figure 6.8 it consists of an illumination system and two stationary cameras which provide the required two 2-D images. This method is inherently more expensive than the other two because it uses 2 cameras but does not require any mechanical movement and, therefore, compared to method 'b' is faster and can provide more accurate measurement of the cameras positions as required for the disparity calculations.

(b) *Disparity method* 2. Use of one imaging device moved to different known positions. This is essentially a cost variation on the method 'a' since, as shown in Figure 6.9, it only differs by the use of a single camera which, to provide images from a different angle, is mechanically moved to a different known position.

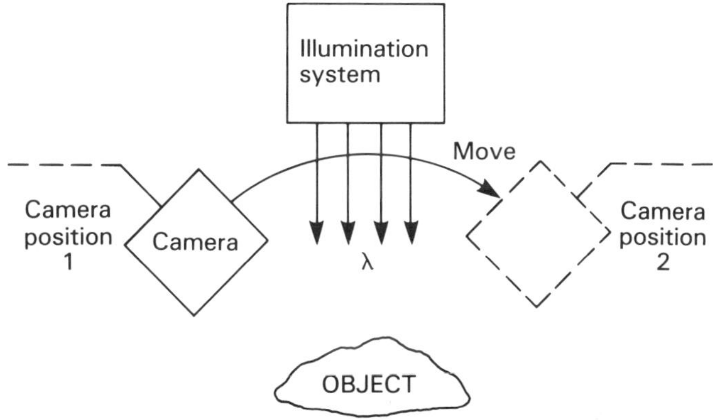

Figure 6.9 Stereo vision, disparity 2 method

(c) *Photometric method.* Use of one stationary imaging device under different lighting conditions. This method relies on maintaining a camera in the same position, thereby avoiding the pixel correspondence problem, and

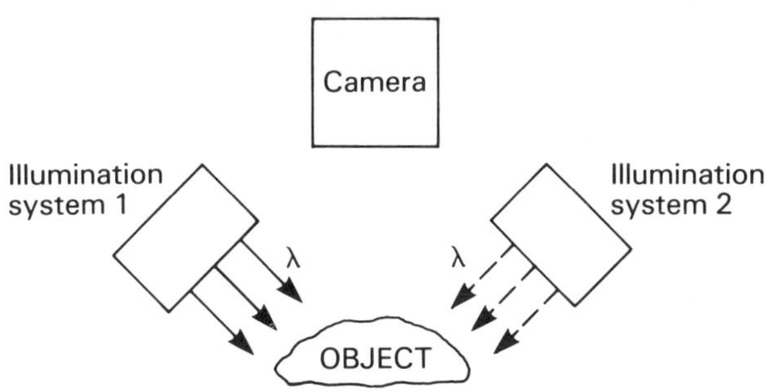

Figure 6.10 Stereo vision, photometric method

obtaining multiple images by changing the illumination conditions. Processing of these images can uniquely determine the object surfaces orientation thus enabling its 3-D mapping (Woodham 1978).

6.5.2 Structured lighting

Structured lighting research has also received wide support both in Europe and in the USA, particularly in the areas of inspection and quality control. This method is based on the idea of using geometric information, encoded in the illumination, to help extract the required geometric information from the object 2-D image. This is achieved by projecting a suitable light pattern (using a high-power light projector or a laser) on the target object and observing the deformations that the object shape produces on the pattern, using a suitable 2-D optical transducer such as a camera and triangulation calculations to obtain the depth map.

The light pattern can be a series of geometrically arranged dots, parallel lines or more simply a sheet of light, depending on the shape of the object and the application. Figure 6.11 shows how a plane of light is projected on to the target object (in this case an optically opaque cube) and produces the 2-D image shown in Figure 6.12. This image, often referred to as the 'raw data', is then processed to extract the 3-D information about the object.

As only that part of the object which is illuminated is sensed by the camera so, in the instance of a single plane of light (as shown in Figures 6.11, 6.12 and 6.13(a)), the image is restricted to an essentially one-dimensional entity, thereby simplifying the pixel correspondence problem. The light plane itself has a known position and every point in the object

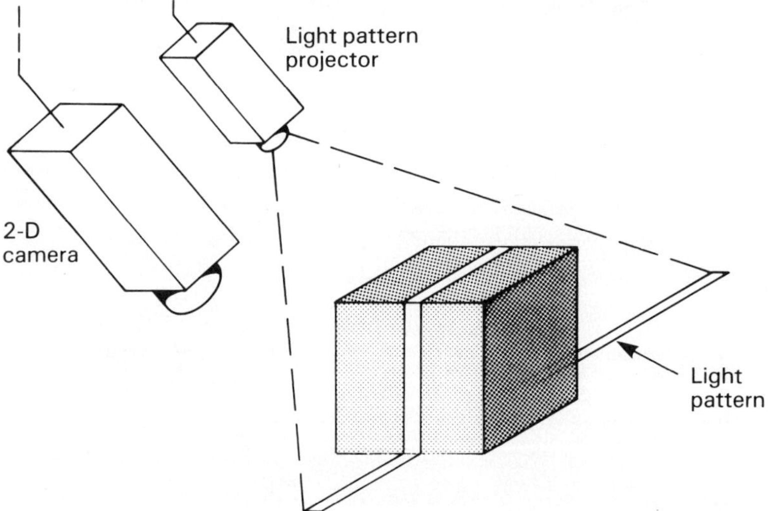

Figure 6.11 Structured lighting generalized diagram (light stripe pattern shown)

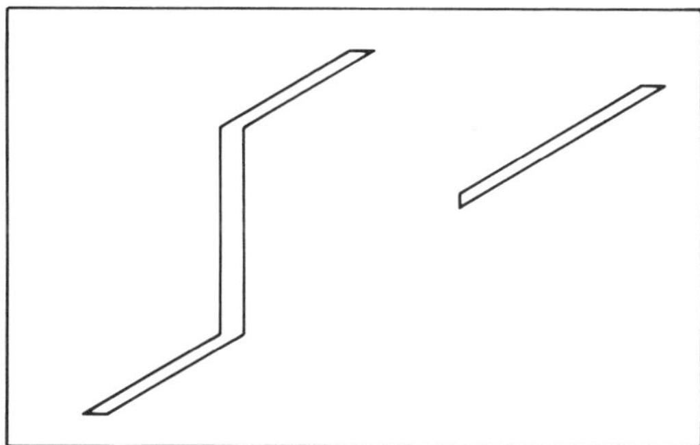

Figure 6.12 Camera output (raw data)

image must also lie on the light plane in the 3-D space. To find out where exactly this point lies (i.e. how far from the projector and the camera) one must consider that the light plane and the line of sight of the camera intersect in just one point (as long as the camera's focal point is not in the light plane). Thus, by computing the intersection of the line of sight with the light plane we can calculate the 3-D position of any object point illuminated by the stripe.

It is important to notice that the only points that can be 'seen' by both the light source and the camera at once can be computed in 3-D. Since the triangulation calculations require a non-zero baseline, the camera cannot be too close to the light source and thus concavities in the scene are potentially difficult to measure, since both the camera and the light source may not be able to see into them at the same time. Another potential problem is created by those object surfaces which are parallel with the light plane since they will have only a relatively small number of lines projected on to them.

This, and similar problems, can be improved by using different light patterns as shown in Figure 6.13(a), (b) and (d). However, the image processing is easier when the light pattern is a single plane of light (Popplestone *et al.*, 1975; Agin, 1972; Sugihara, 1977) than when it is a series of geometrically arranged dots or stripes, in view of the higher visual data content provided by the two latter techniques.

Robot vision based on structured lighting has recently produced some industrial applications worthy of note both in the USA and the UK (Nelson, 1984; Meta Machines, 1985; Edling, 1986).

6.5.3 Dynamic focusing

Dynamic focusing has not received as much attention as the other two indirect 3-D vision methods but represents an important alternative particularly in terms of price and speed of operation. This technique relies upon correlating the signature between two linear photodiode arrays to obtain the range information (Stauffer and Wilwerding 1984) and is therefore in-

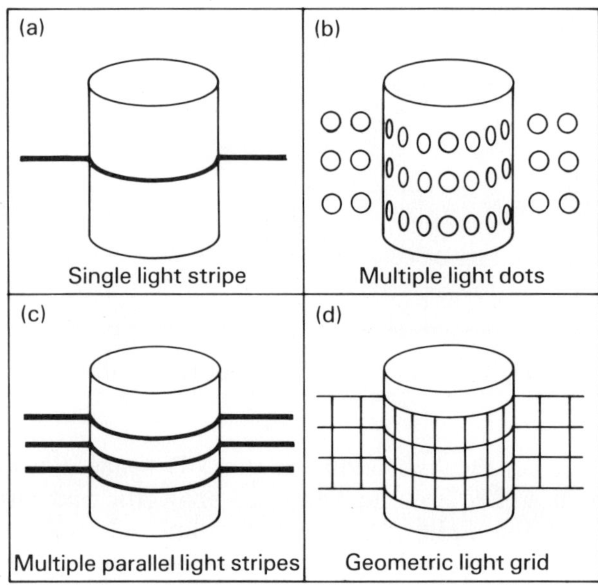

Figure 6.13 Different types of structured light patterns

herently less flexible than structured lighting or stereo vision (it relies on conveyor belt movement to provide the third dimensional axis) but is eminently suitable for object dimensioning, edge detection and tracking.

Vision sensors based on dynamic focusing make use of the automatic focusing technology first developed in the mid-1970s for the 35 mm SLR and video camera markets. They sense the relative position of the plane of focus by analysing the image phase shift which occurs when a picture is not in focus.

The principle of operation is as follows: when an image is in focus, all the light arriving at a single point on the image plane is produced by a corresponding single point in the scene. That is to say all the light collected by the lens from this single point in the scene is focused on the corresponding single point (and this one only) on the image plane. The whole lens, viewed from a point in the image, must therefore appear to be one uniform colour and brightness, like the corresponding point in the scene. It thus follows that, if every point in the image were divided into two halves, they would both have the same values of colour and brightness (i.e. their 'signature' would be the same). When the scene image is not in focus the same two halves of the image point would not have the same values of colour and brightness (i.e. a different 'signature'), the degree of difference providing a measure of how far out of focus the scene image is.

This is the same mechanism upon which dynamic focusing vision sensors, such as the Honeywell HDS-23, are based. The front-end optical transducer, as shown in Figure 6.14(a), is a single row of twenty-three light

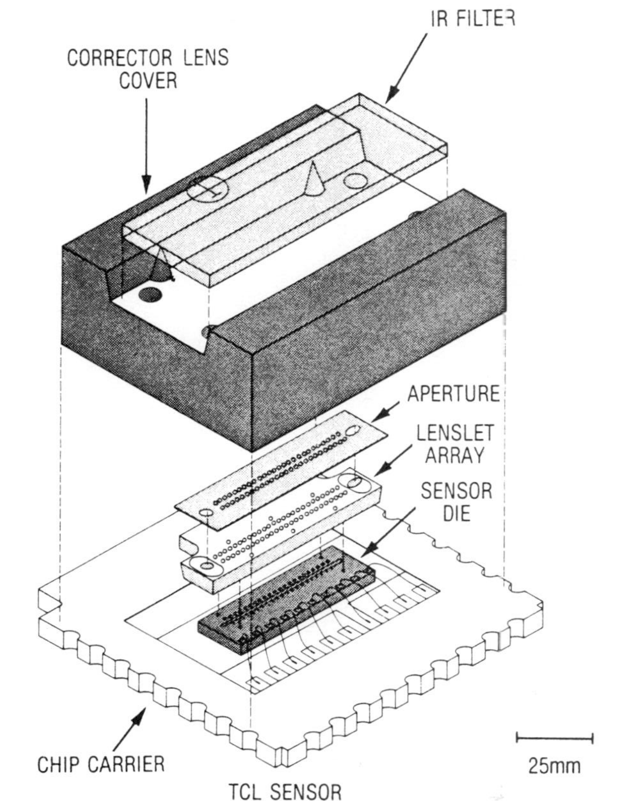

Figure 6.14(a) Exploded view of Honeywell-HD23 dynamic-focusing vision sensor (courtesy of Honeywell Inc.)

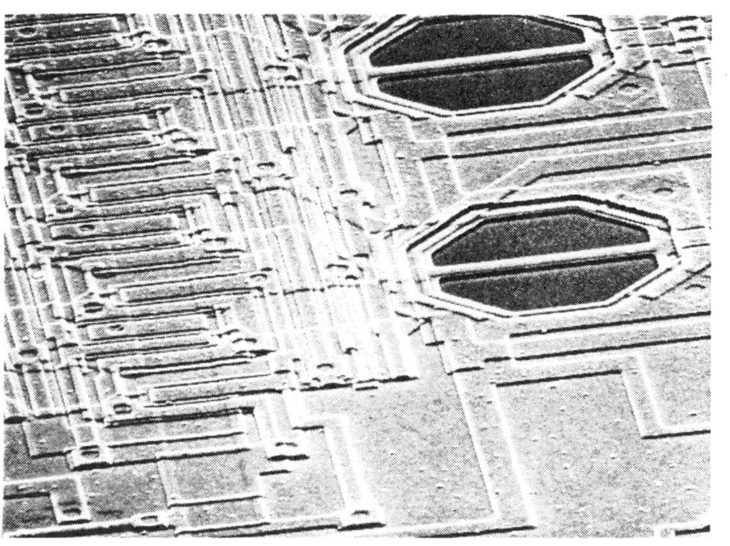

Figure 6.14(b) Electron microscope photograph of two light detector cells (divided clearly into two halves) and the adjacent CCD shift register array (courtesy of Honeywell Inc.)

cells which are similar to the cells in a solid-state camera, but with two major differences: each of the cells has its own miniature lens, about the diameter of human hair, and is made up of two halves, a right half and a left half. Thus, each of the cells in the HDS-23 sensor can look at the surface of the lens from a single point on the image plane and determine if the light brightness from the right half of the lens is the same as that from the left half. Local intelligence within the sensor then computes a 'signature' of the scene on the image plane, as seen by the twenty three left and right light cells, and determines if these two signatures are the same (image in focus) or are not the same (image not in focus).

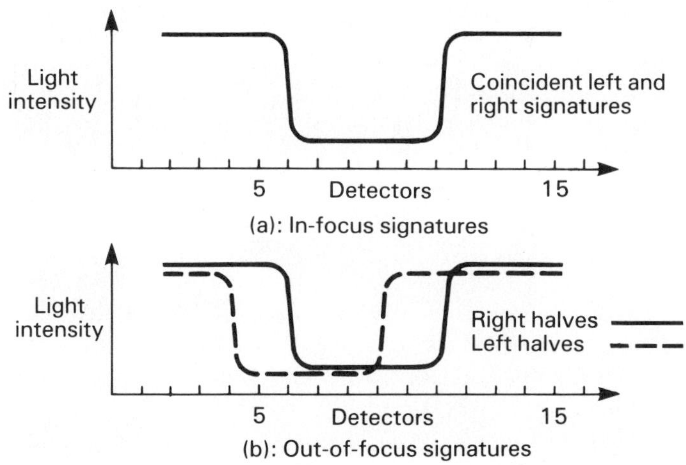

Figure 6.15 Dynamic focusing image signatures (courtesy of Honeywell Inc.)

It turns out, as shown in Figure 6.15, that for a 2-D scene, for instance a black square on a white background, the out-of-focus signatures are actually the same but shifted in phase. By computer processing the direction and the magnitude of the signature phase shift, the sensor can determine the

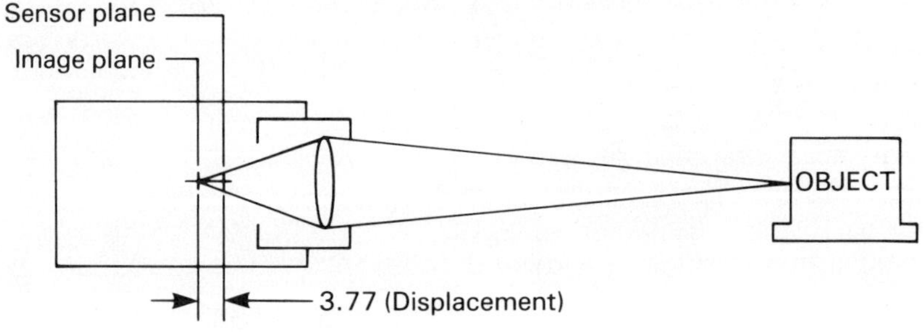

Figure 6.16 Dynamic focusing sensor optics (courtesy of Honeywell Inc.)

magnitude and the direction of the image plane displacement from the focus condition and, as shown in Figure 6.16, determine the object distance (Iversen, 1983).

Honeywell-Visitronics has recently commercially released a range of robotic sensors based on the dynamic focusing technique (Honeywell-Visitronics, 1984).

6.5.4 Range-measuring sensors

Direct range measurements offer considerable advantages over the triangulation methods both in terms of speed of operation and data-processing efficiency (Jarvis, 1982) but tend to have a lower image resolution, particularly along the x–y plane, and to suffer from the wide dynamic range of the returned signal. This latter point leads to the requirement for sophisticated detection circuits, which adds to complexity and, to a lesser extent, cost but which helps to shift the sensor design emphasis towards hardware rather than software as in the case of triangulation-based 3-D vision sensors (such as stereo and structured-lighting ones) thereby allowing for the potentially higher speed of operation (e.g. the rangepics are measured not calculated thus avoiding a major step of the data manipulation software).

The majority of the research work in this field has been concentrated on scanning laser rangefinders (Jarvis, 1983; Page *et al.*, 1983; Nimrod *et al.*, 1982; Nitzan, 1977; Lewis and Johnson, 1977), static rangefinders (Ruocco, 1985; Kanade and Somner, 1983) and ultrasonic rangefinders (Polaroid Co.)

Scanning optical rangefinders
In the area of optical 3-D devices, scanning laser rangefinder research has received wide attention during the last decade and spanned the field of robotics as well as those of aeronautics and military applications (Johnson, 1977; Nitzan, 1977; Page *et al.*, 1983; Nimrod *et al.*, 1982; Kulyasov *et al.*, 1973; Matthys *et al.*, 1980; Kaisto, 1983).

As shown in Figure 6.17, the technique relies upon mechanical scanning of the target object by a laser beam whose reflected light is processed to obtain the range data; the resulting depth map has, potentially, a resolution of less than a millimetre. This makes the scanning laser rangefinder an important 3-D vision sensor but its relatively high price and inherent mechanical design can be a drawback in applications where low cost and/or rugged sensors are required. This is thought to be one of the underlying reasons why scanning laser rangefinders have yet to find wide application in the robotics industry.

Figure 6.18 gives an example of the typical output provided by a laser rangefinder as measured by Lewis and Johnson in 1977, the lighter regions being closer to the sensor. Figure 6.18(a) shows a range-picture of a chair, whereas Figure 6.18(b) shows a range-picture of the same chair with a person sitting in it.

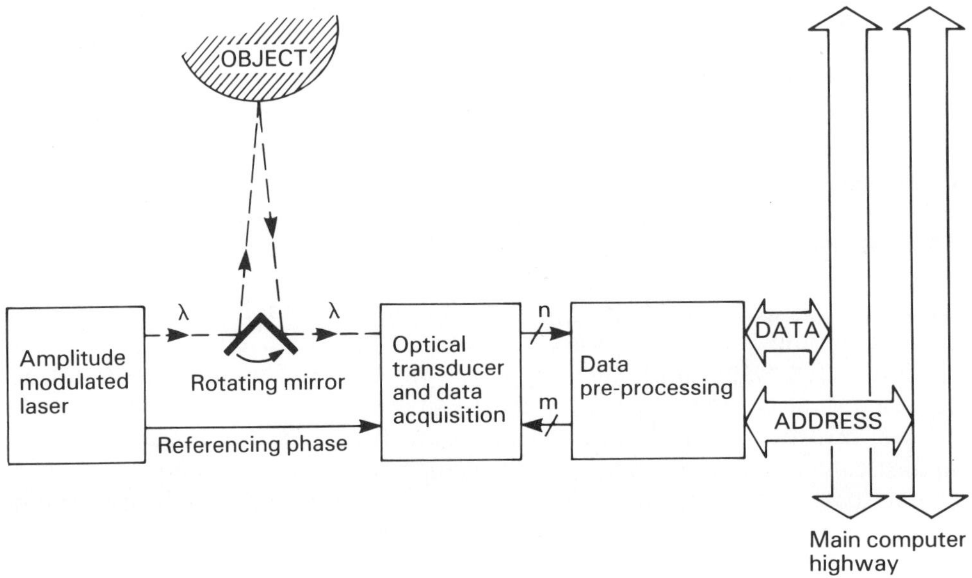

Figure 6.17 Generalized diagram of scanning laser range finder

Static rangefinders

Static LED arrays offer two main advantages over scanning laser rangefinders: they have no moving parts and employ cheaper light sources. They do, however, suffer from a lower optical launch power and, potentially, also a lower x–y resolution which tends to limit their use to low-cost, short-range applications. Thus they are eminently suitable for eye-in-hand robot vision applications particularly in the case of multisensor robot systems where they can fill the natural gap between the visual sensing of an overhead camera and the pressure sensing of a tactile sensor.

Figure 6.18 Typical output of scanning laser rangefinder (after Lewis and Johnson)

The principle of operation of a static LED array rangefinder is essentially the same as that of the scanning laser device, that is each LED output is focused on to the target object whose reflected light is processed to provide the range data, but the scan is achieved electronically by multiplexing the drive to the LED matrix so that only one LED is 'on' at any point in time, thereby avoiding the need for a potentially troublesome mechanical scan. Two main techniques have been developed to achieve this goal—an $x-y$ LED array sensor and a circular LED array sensor (Ruocco, 1986; Kanade and Somner, 1983).

A *2-D LED array sensor* has been developed by Ruocco at Middlesex Polytechnic based on the phase measurement principle. As shown in Figure 6.19 this sensor has the light-emitting diodes arranged in a 2-D array with the same $x-y$ resolution as the intended object range image.

The principle of operation, as illustrated in Figures 6.20 and 6.21, is as follows: a single LED is modulated at an appropriate frequency and is focused on to the target object. The subsequent reflected/scattered optical flux ϕ_s is coupled to the secondary detectors whose output signal V_s is compared with the reference signal V_r provided by the primary detectors which receive light only (and directly) from the LED array. The phase difference between these two signals provides a measure of the object distance, as shown by eqns (6.1) and (6.2):

$$\text{signal time of flight} = t_d = \frac{2 \cdot d}{c} \tag{6.1}$$

$$\text{signal phase lag} = \Delta\varphi = \omega_M \cdot t_d = \frac{2 \cdot \omega_M \cdot t_d}{c} = k \cdot d \tag{6.2}$$

where ω_M is the LED modulating frequency and c the speed of light.

The procedure is repeated for all LEDs in the array and a 3-D scan of the target object is thus achieved.

Local intelligence provides a degree of 3-D image pre-processing such as 'closest object point' and 'holdsites location' which helps to reduce communication with and processing times by the main robot computer, a desirable goal in most applications such as obstacle avoidance and pick-and-place operations.

The features of such a sensor are low cost, high speed and medium $x-y$ resolution. This latter furthermore increases as the sensor approaches the target object thus making it ideal for eye-in-hand operation and integration within a robot Multisensory Feedback System. Embryonic examples of such systems (like the one based on an overhead camera, a 3-D eye-in-hand vision sensor and a tactile sensor) have been shown to provide a considerable increase in the robot flexibility of operation and are the subject of wide ranging research interest (Ruocco and Seals, 1986; Ruocco, 1986, Dillman 1982; Van Gerwen and Vleeskens, 1985; Andre, 1985).

A *circular LED array sensor* has been developed by Prof. Kanade at

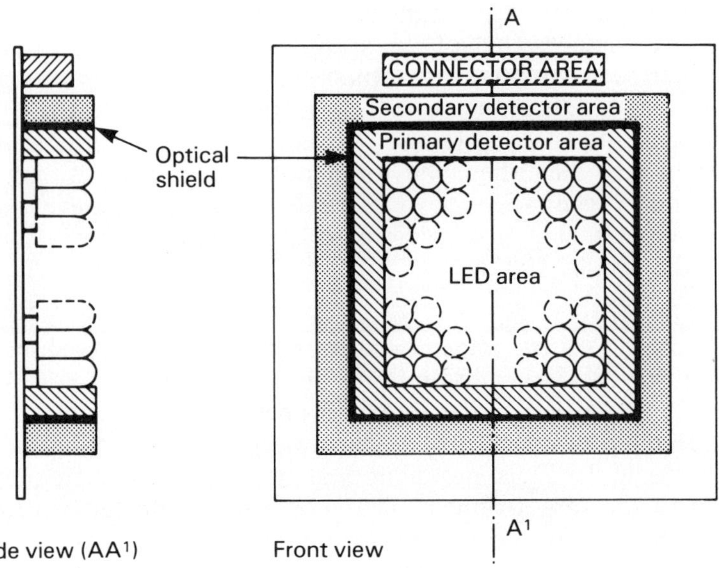

Figure 6.19 Orthogonal projections of 3-D vision sensor (front end only)

Carnegie-Mellon University based on a dedicated light spot-sensing device. The sensor front-end, as shown in Figure 6.22, consists of a circular array of LEDs, an objective lens and a dedicated optical transducer acting as a light spot position detector.

The principle of operation is as follows: the LEDs are aligned so that all the light beams cross the optical axis at the same point, thus forming an

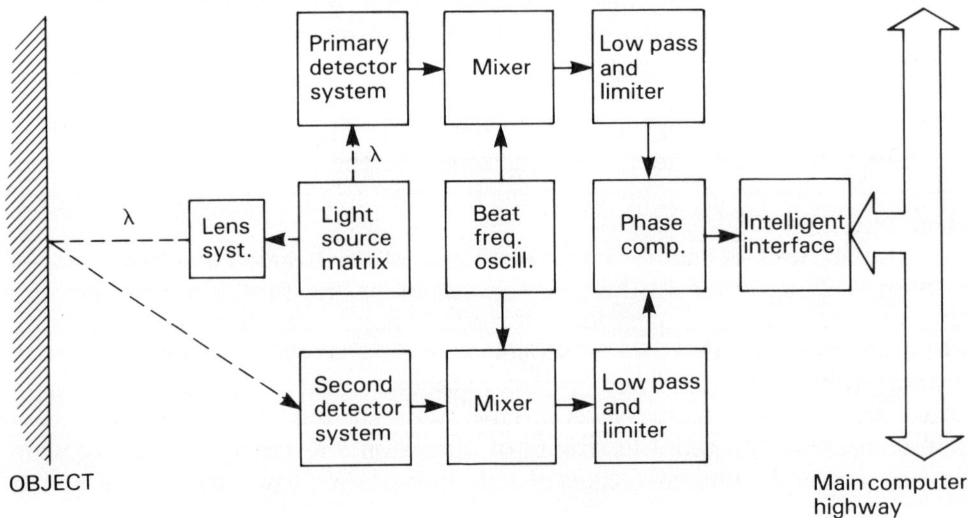

Figure 6.20 Simplified block diagram of 3-D vision sensor

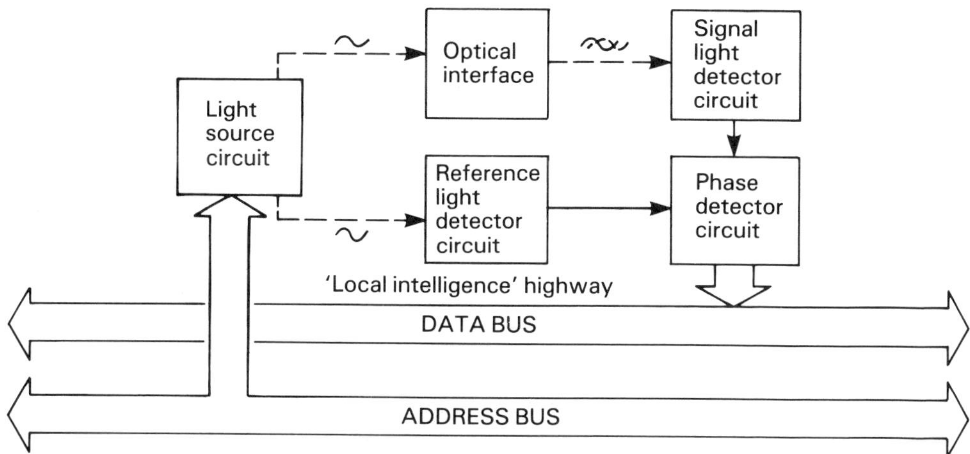

Figure 6.21 Functional diagram of 3-D vision sensor (front-end data acquisition section)

optical cone, the tip of which is placed at approximately the centre of operation of the 3-D vision sensor. The objective lens and the optical transducer form a camera arrangement so that when an object intercepts the light beam the resultant light spot is imaged on to the optical transducer, a planar PIN photodiode with homogeneous resistive sheets on both sides (the Hamamatsu S1300) which, as shown in Figure 6.23, can measure the photocurrents produced by the light spot on an x–y plane.

These values are then used by the local intelligence to evaluate the light

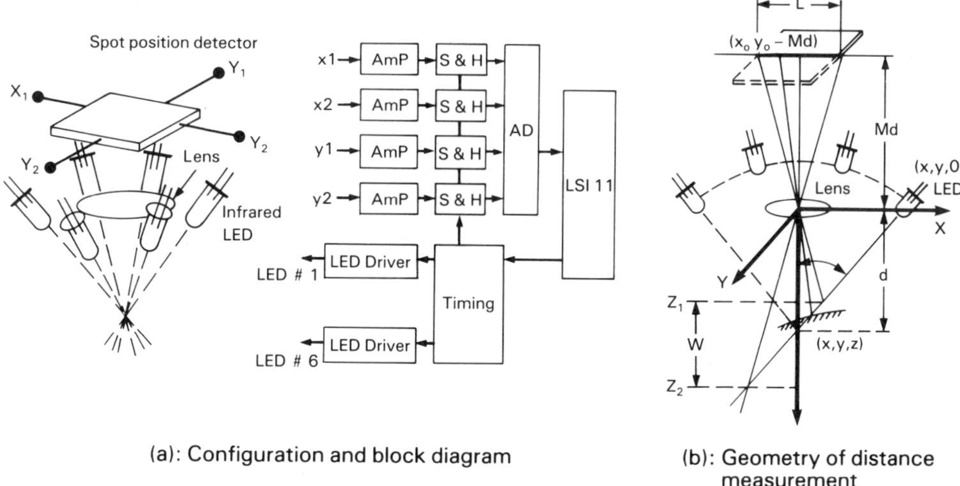

Figure 6.22 Rangefinder based on circular LED array (after Kanade and Somner, courtesy of SPIE 1983)

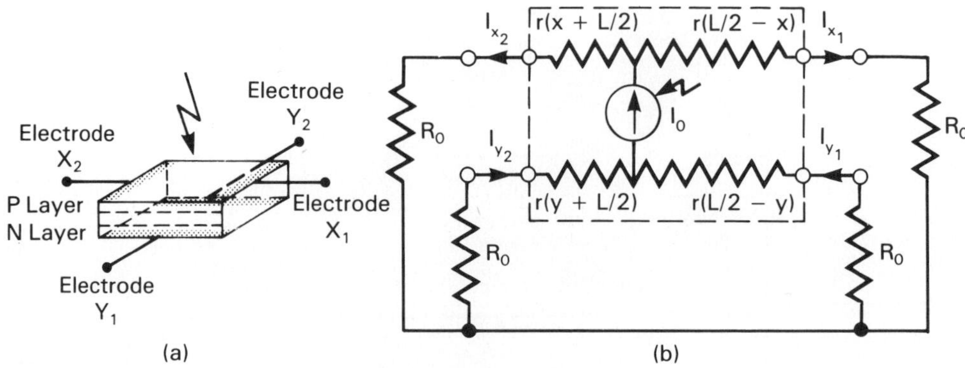

Figure 6.23 Optical transducer functional diagram (after Kanade and Somner, courtesy of SPIE 1983)

spot position according to eqns (6.1) and (6.2), where r is the resistance of the homogeneous resistive sheet, L is the length of the transducer and I is the electrode current:

$$X_c = \frac{I_{X1} - I_{X2}}{I_{X1} + I_{X2}} = \frac{2rx_c}{rL2R_0} \tag{6.1}$$

$$Y_c = \frac{I_{Y1} - I_{Y2}}{I_{Y1} + I_{Y2}} = \frac{2ry_c}{rL2R_0} \tag{6.2}$$

Knowledge of the light spot x–y position, the camera optics and the trajectory of the light beam allows triangulation calculations to be performed, in accordance with eqn (6.3), to find the object point coordinates in 3-D. From the measurement of multiple close object points, the surface orientation can be calculated.

$$Z = \frac{d}{1 - \left(\frac{X_c}{dM \tan \theta}\right)} \tag{6.3}$$

The features of this sensor are: the simplicity of the operating principle, fast speed (it can generate up to 1000 range measurements per second) and precision (0.07 mm for distance and 1.5° for surface orientation). Its drawbacks, compared with the 2-D linear LED array technique, are: the larger size (Figure 6.24 shows the typical size for a six LEDs sensor), and the smaller range of operation (4–5 cm compared with 10–80 cm typically), both thought to be caused indirectly by the inherent construction requirement that the LEDs be angled with respect to the transducer optical axis.

Development of this sensor is still proceeding, with Prof. Kanade and his team currently working on a prototype that enables the use of LED

Figure 6.24 Photo of the prototype proximity sensor; 8 cm (diameter) × 10 cm (length) (after Kanade and Somner, courtesy of SPIE 1983)

modulation (to allow operation under ambient lighting conditions), fibre-optic cables and laser diodes (to increase the optical incident power) and double optical cones (to eliminate the singularity of 'plane fitting').

Another example of an optical range sensor capable of providing a 2-D or 3-D image (depending on the light source configuration used) is the one proposed by Page *et al.*, at Lanchester Polytechnic [Page *et al.*, 1981, 1983). Like the one proposed by Ruocco (1986), this sensor is also based on measuring the phase difference between the launched and the returned optical signal and is therefore based on the same theory. The main difference lies in the use of laser diodes in preference to LEDs. This choice affords higher modulation frequency and therefore a potentially higher range measurement resolution. Unfortunately the higher unit cost, larger parametric variations and more complex drive circuit requirements of laser diodes also limit their use to individual light sources (in the same fashion as scanning laser rangefinders) or as 1-D arrays which then requires knowledge of the object movement, as in the case of conveyer belt use, to produce a 3-D image.

One instance of a research device successfully developed and marketed and now commercially available is the single point optical rangefinder based on triangulation (Electronic Automation, 1984). This device was first proposed in the late seventies and subsequently developed by several researchers (Okada, 1978, 1982; Tanwar, 1984; Edling and Porsander, 1984) to yield the present day product. The principle of operation is illustrated in Figure 6.25.

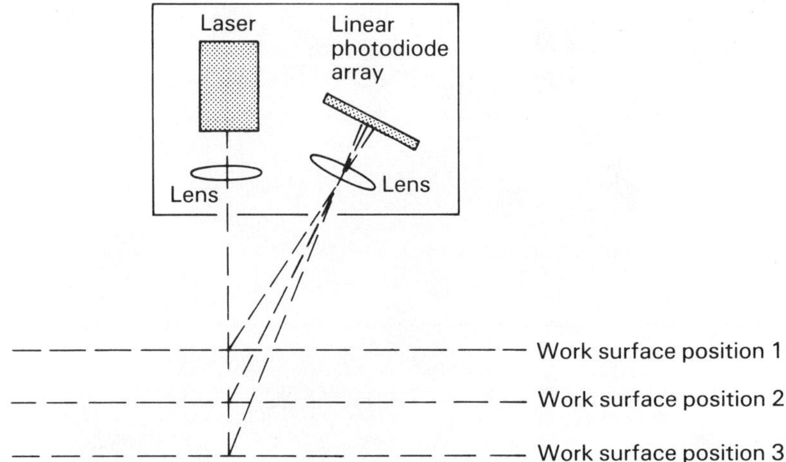

Figure 6.25 Principle of operation of single point laser rangefinder based on triangulation

These, however, are single point devices, that is they are only capable of providing the distance measurement to one point on the object surface and, because of their size, are unsuitable for mounting as an $x-y$ array which would be necessary in order to provide a 3-D image. This sensor geometry does, however, allow high precision single point range measurements (for instance, Tanwar quoted a precision of 0.01 micrometers for polished surfaces) albeit within a small range of operation (±75 micrometers for the Tanwar sensor).

Ultrasonic rangefinders
Ultrasonic imaging has come to the fore in recent years as an important alternative to X-rays in the field of medicine (Karrer, 1983; Waaft and Gramiak, 1976). In spite of this success, however, the ultrasonic technique has yet to provide any new solutions to the problem of rangefinding in the field of robotics. This is thought to be due mainly to the large mismatch between the impedance of the ceramic transducers and that of the air medium, as well as the wide emission angle of most ultrasonic transducers.

An interesting research program is, however, being carried out at the University of Canterbury (NZ) by Prof. L. Kay and his team on the development of an ultrasonic sensor suitable for robotic applications.

The present prototype is only a single point rangefinder but the small size of the transducers indicate the possible expansion to a limited 2-D linear array.

Because of their wide angle of operation, single point ultrasonic rangefinders have, in fact, useful applications in obstacle detection in such fields as mobile robotics (Seals, 1984). The best known single point distance sensor currently on the market is the Polaroid ultrasonic sensor developed primarily for photographic camera applications. Since the Polaroid sensor is, in principle, very similar to others in this field, it will be used as the mean of explanation. The basic structure of the front end ultrasonic transducer is shown in Figure 6.26.

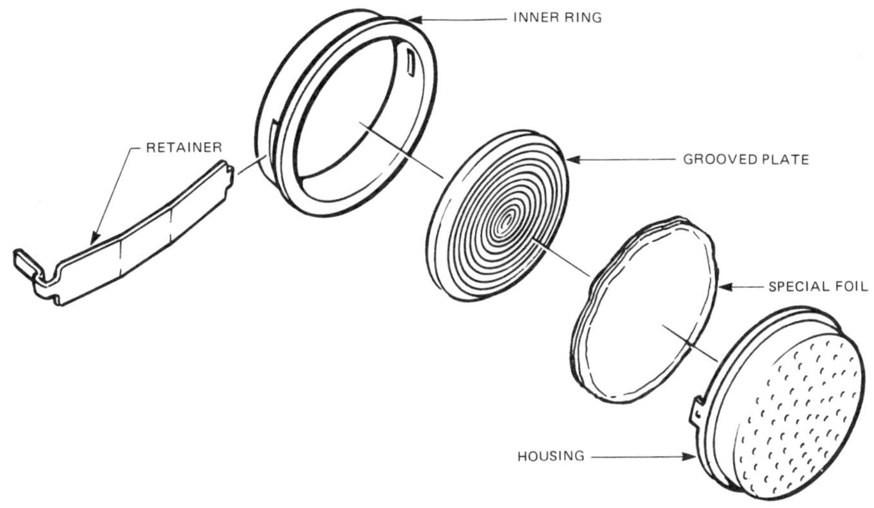

Figure 6.26 Exploded view of an ultrasonic transducer (courtesty of Polaroid)

The principle of operation of an ultrasonic sensor is to measure the time delay t between the transmitted and reflected sound pulses which, assuming a constant velocity v for sound propagation, is related to the obstacle distance d by the simple formula $d = vt$.

Figure 6.27 illustrates the main steps of the ultrasonic distance measuring method. A sound pulse is produced by the ultrasonic transducer (pulse length approximately 1 ms and frequency spectrum from 50 kHz to, typically, 60 kHz) which, after a time delay proportional to the object distance from the sensor, also receives the reflected sound pulse or 'echo'. The hardwired local intelligence then processes these two signals (emitted and reflected pulses) and calculates the obstacle distance from the sensor.

Since the received 'echo' could have been produced by any suitable object within the incident sound pulse 'cone', whose output beam pattern is shown in Figure 6.28, a scan of the required area needs to be carried out in order to obtain a range image if required, as in the case of obstacle recognition. This can be achieved either electronically, by using an array of ultrasonic transducers, or mechanically, by moving the single transducer in a suitable pattern. It is interesting to note that, since the transmitting cone changes with frequency, a small degree of electronic 'scan' is achieved by

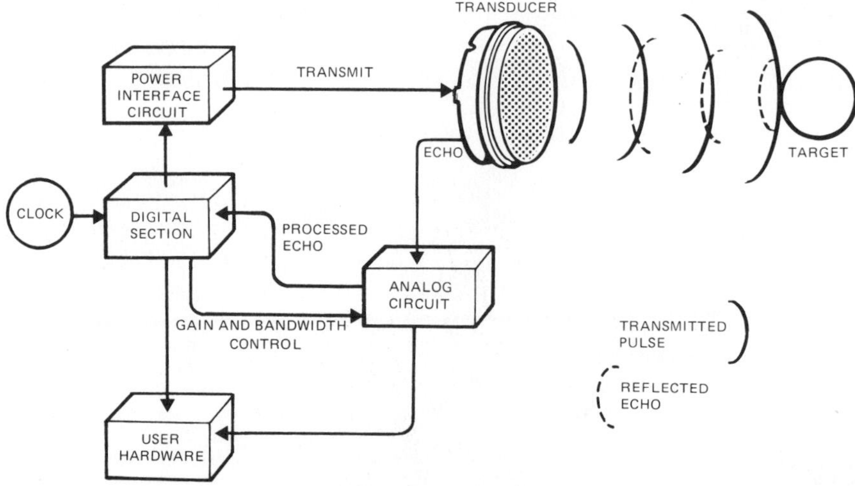

Figure 6.27 Ultrasonic sensor functional block diagram (courtesy of Polaroid)

incorporating a frequency sweep within the emitted sound pulse; a typical frequency range would be 50–100 kHz.

Another useful outcome of the frequency sweeping (also achieved by using multiple discrete frequency values, as in the case of the aforementioned Polaroid distance sensor) is that there is less chance of one object being particularly transparent (or absorbent) to the transmitted frequency thereby

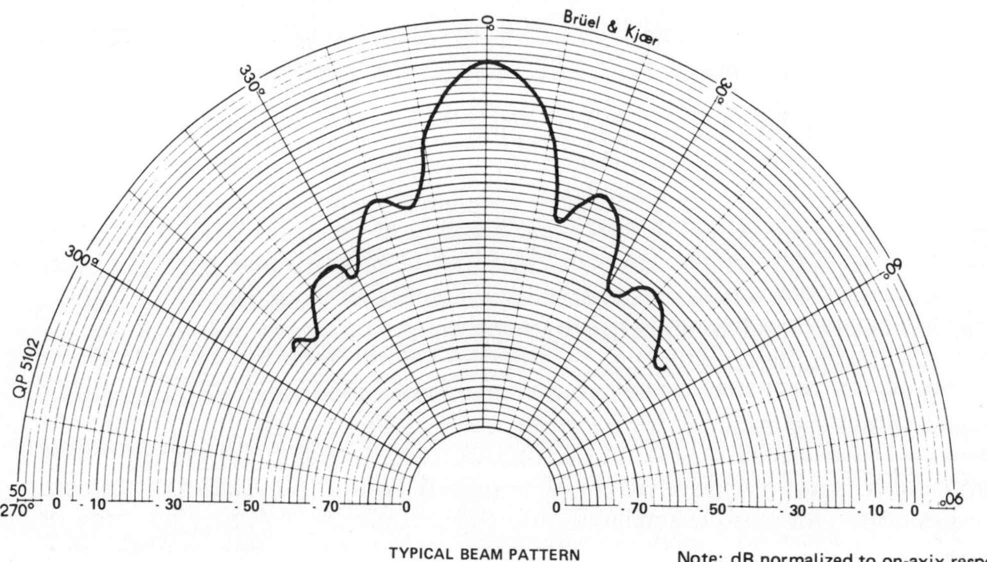

Figure 6.28 Typical ultrasonic transducer output (@ 50 kHz) (courtesy of Polaroid)

causing faulty operation, since this object would not be 'detected' by the ultrasonic sensor.

6.6 Interfacing of vision sensors

Vision sensors are a relatively new technology with few common features. Interfacing these sensors therefore requires an in depth knowledge of the individual sensor being used and commands strict adherence to the manufacturer or, more generally the research establishment, data sheets.

Although internally very complex these sensors, however, usually require few external control signals and, as a general guideline, need the less handshaking the more local intelligence they have. An example of basic interfacing is shown in Figure 6.29 for the Fairchild ISCAN 1-D vision sensor as extracted from the manufacturer CCD catalog (Fairchild, 1984).

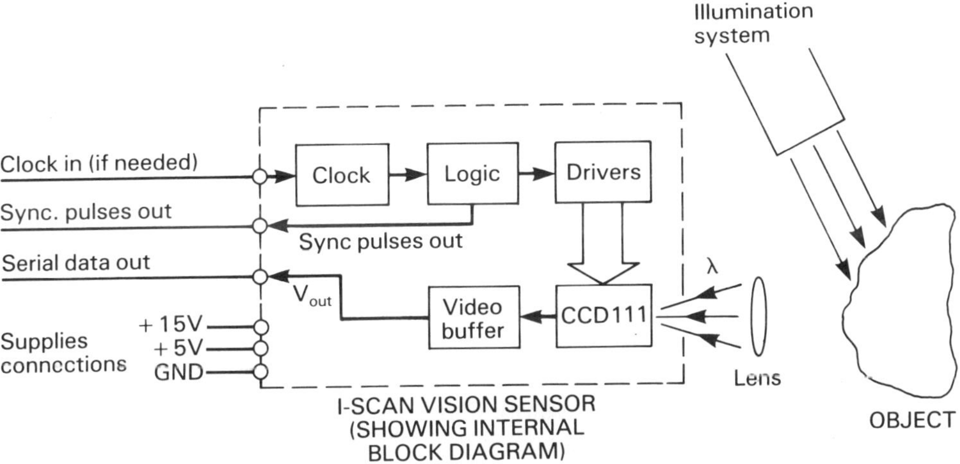

Figure 6.29 Interfacing the I-SCAN vision sensor (based on the CCD111 linescan camera; courtesy of Fairchild, 1984)

This sensor is based on the 1-D optical array transducer CCD111 (i.e. a linescan camera) and generates internally all the necessary waveforms shown in Figure 6.30, thus requiring only an external clock for synchronization purposes, an exposure signal (if the image exposure needs to be under software control) and the supply lines in order to provide the output image on a serial line (for further information on CCD cameras, see Section 3.5.2.3).

Another example of specific yet simple vision sensor interfacing is provided by the EV1 sensor manufactured by Micro Robotics. This device is based on the IS-32 DRAM chip and, once again, generates internally all the necessary waveforms thus requiring only 3 signals: a clock signal of between

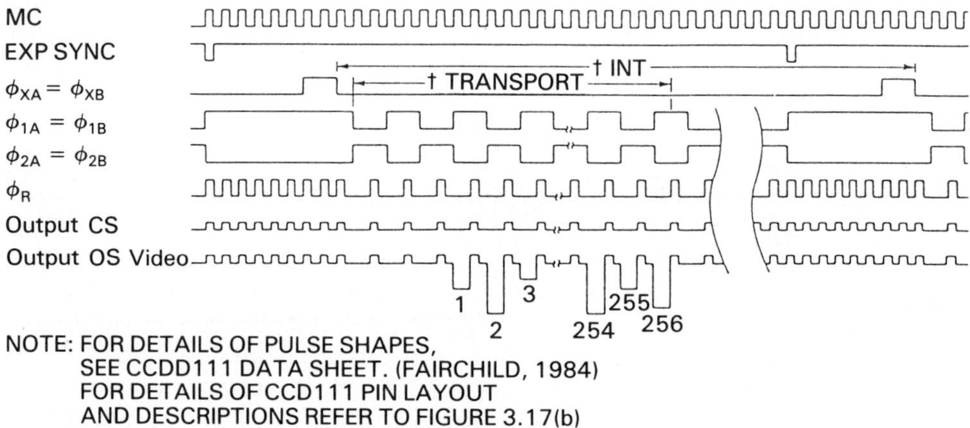

Figure 6.30 I-SCAN vision sensor timing diagram (courtesy of Fairchild, 1984)

1 and 10 μS, a reset signal and a separate logic signal to determine which half of the DRAM is being used (this latter is only needed because of the particular optical arrangement of the IS-32 chip). Figure 6.31 shows the interfacing of the EV1 to the AppleIIe microcomputer as reported by Marr (1986); for further references on the DRAM, see Section 3.5.2.4.

6.7 Conclusions

Vision sensors provide the most powerful yet flexible environmental sensing available for robot systems development. Tough 2-D vision is still used in industry, 3-D vision sensors are currently being developed in the laboratories throughout the world and will soon enable the full exploitation of the second and eventually third generation of robots.

6.8 Revision questions

(a) Describe, using examples, the differences between 'through' and 'reflected' illumination and their effect on image acquisition.
(b) Explain the principle of operation of an ultrasonic rangefinder. Why can this principle not be easily applied to optical rangefinders for measuring distances shorter than 1 metre?
(c) Discuss the use of vision sensors within a robot system designed to pick random parts from a bin and propose solutions to any problems you may encounter.

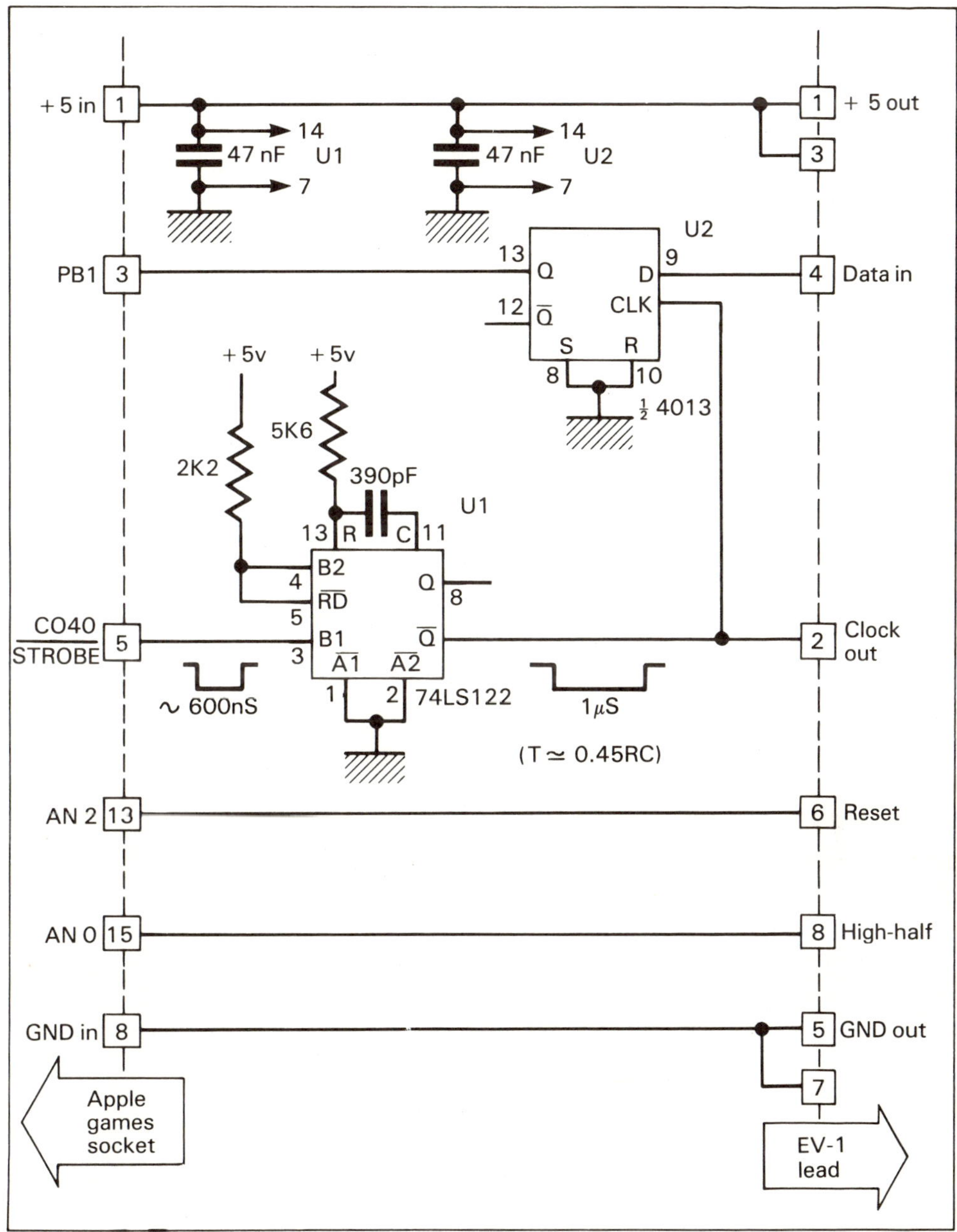

Figure 6.31 Interfacing the EV1 vision sensor (based on the IS-32 DRAM camera) to the Apple IIe microcomputer (after Marr, 1986 and courtesy of Apple User)

6.8 Further reading material

Batchelor, B. G., Hill, D. A. and Hodgson, D. C. (1984) *Automated Visual Inspection,* London, IFS.

Chappell, A. (ed.) (1976) *Optoelectronics—Theory and Practice,* Texas Instruments Ltd.

McCloy, D. and Harris, M. (1986) *Robotics: An Introduction,* Milton Keynes, Open University Press.

Pugh, A. (ed.) (1983) *Robotic Technology,* Hitchin, Peter Peregrinus.

Snyder, W. E. (1985) *Industrial Robots: Computer Interfacing and Control,* Englewood Cliffs, Prentice-Hall.

Chapter 7

Tactile sensors

7.1 Overview

Tactile sensors, like vision systems, have been the object of intense research throughout the world. Indeed it is increasingly evident that the limitations suffered by vision systems could be overcome by the cooperative use of tactile sensors and that Robot Multisensory Feedback Systems, comprising at least one 2-D vision system, a rangefinder and a tactile sensor, are a good base for the development of future generations of robots (Ruocco and Seals, 1986).

As mentioned earlier in Chapter 4, tactile sensors are based on direct measurement force transducers, that is devices which measure the mechanical deformation produced by a force acting directly on the transducer itself. This principle was illustrated in Figure 4.1 and is shown again below in Figure 7.1. Tactile sensors are based on 2-D arrays of such measuring 'cells'. Like their human analogue, in fact, robotic tactile sensors work by measuring the contact pressure between the object surface and the robot gripper; by dividing the contact area into an array of several measurement points (typically 16×16), a 2-D image of the contact pressure can be mapped, thereby producing a 3-D view, though partial, of the object being manipulated by the robot gripper.

It should be noted at this point that, following a generally accepted convention, 'tactile sensing' is herewith defined as the continuous measurement of contact pressure within an array of so called 'tactels', or tactile elements, as distinct from 'touch sensing' which is defined as a single contact pressure measurement, as was the case with the force transducers discribed in Chapter 4.

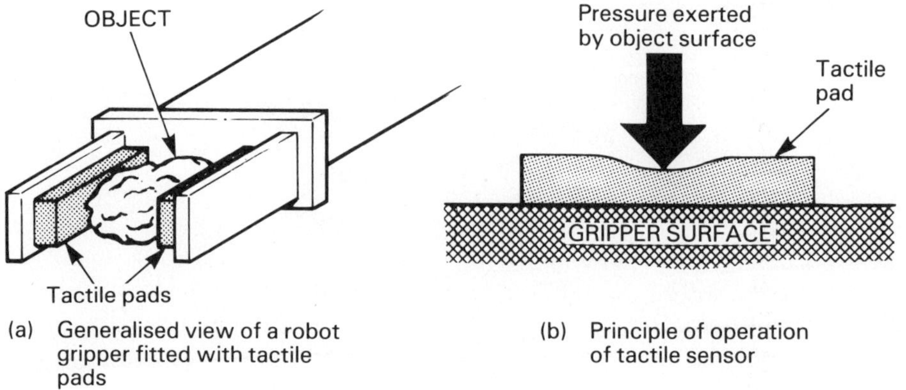

Figure 7.1

7.2 Touch sensing

Touch sensing can thus be accomplished by touch probes which, although can provide only single-point measurements, are often adequate for sensing the mere presence of an object and/or the force exerted on its surface. An example of such an application is 'contour following', where the touch probe is used to guide the robot gripper over the object surface in order to allow some other critical robot operation, such as deburring, welding or even an inspection task like the ultrasonic scan used to detect flaws beneath the object surface, as shown in Figure 7.2:

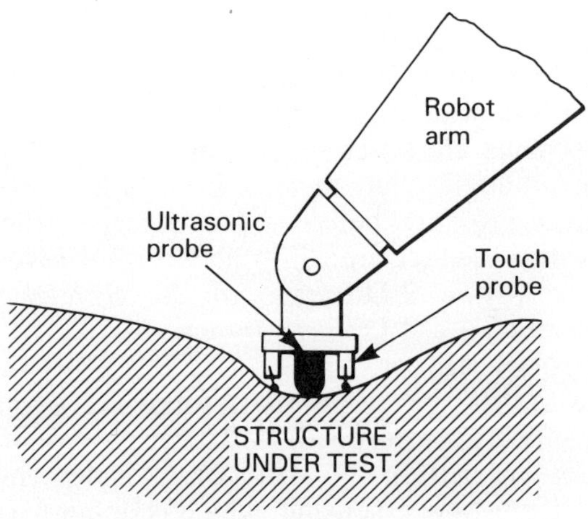

Figure 7.2 Example of contour following using touch probes

7.3 Tactile sensing

Tactile sensing allows continous measurement of the pressure distribution across the tactels array. As well as providing a pressure image of the object at any point in time, the tactile sensor can hence also allow the detection of slippage, another important parameter in the field of object manipulation. This is achieved by looking at the derivative of the pressure images, a task which in practice is achieved by simple consecutive image comparison. Slippage detection is particularly important when handling fragile objects since the forces exerted on their surface must be minimal, in other words the detection of slippage is required by the robot computer to allow the gripper to apply just enough pressure to prevent dropping the object in question.

There are three main types of tactile technology—resistive, capacitive and optical.

7.3.1 Resistive tactile sensors

Resistive devices are the most popular tactile sensors at present, because of their higher sensitivity and x–y resolution. They can be based on a variety of resistive compounds such as carbon fibres, conductive rubbers and various specially developed media impregnated with conductive dopants. Although they all exhibit the required property of a change in resistance in response to an applied pressure, the last category has achieved the highest popularity because of its lower hysteresis and larger dynamic range.

There are two main ways to measure the resistance change—in line with the applied force vector, hence measuring vertical resistance changes, and perpendicular to the applied force vector, hence measuring horizontal resistance changes. An example of a tactile sensor which measures resistance change on a vertical plane is shown in Figure 7.3(a). The resistive mat used

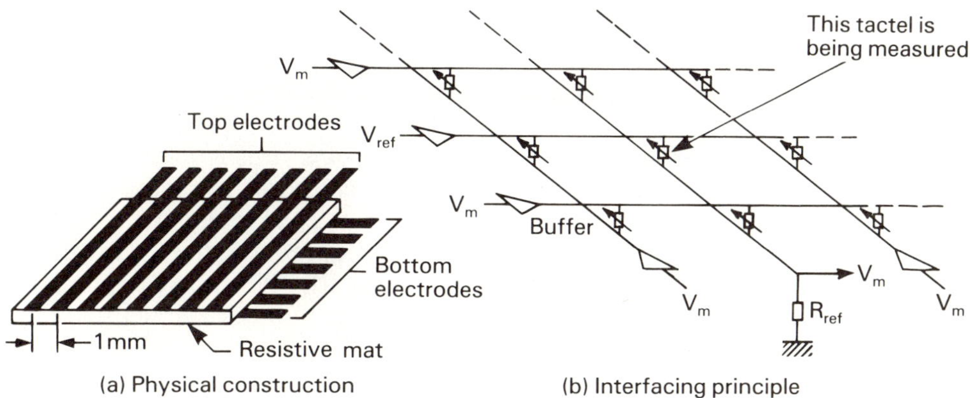

Figure 7.3 Piezoresistive tactile sensor (after Robertson and Walkden, courtesy of SPIE, 1983)

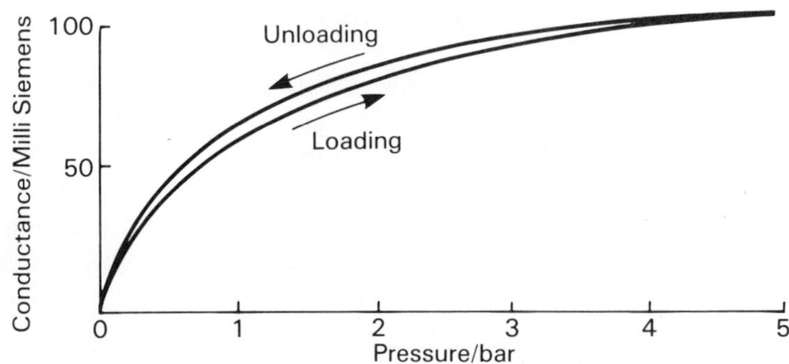

Figure 7.4 Piezoresistive tactile sensor characteristic (after Robertson and Walkden, courtesy of SPIE, 1983)

in this sensor is a piezoelectric material whose sensitivity curve is shown in Figure 7.4 which highlights the low hysteresis of this type of device.

The advantage of such a construction is the higher x–y tactel resolution achieved by having the two electrode patterns on separate parallel planes. The consequent drawbacks are the higher crosstalk due to the capacitive effect, which can however be compensated for by the interfacing electronics as previously shown in Figure 7.3(b), and the shorter life of the top electrode, which needs to be deformed mechanically by the applied force for each measurement. A typical output of the tactile sensor thus described is represented by Figure 7.5 which shows the pressure image of a rectangular object on an isometric display.

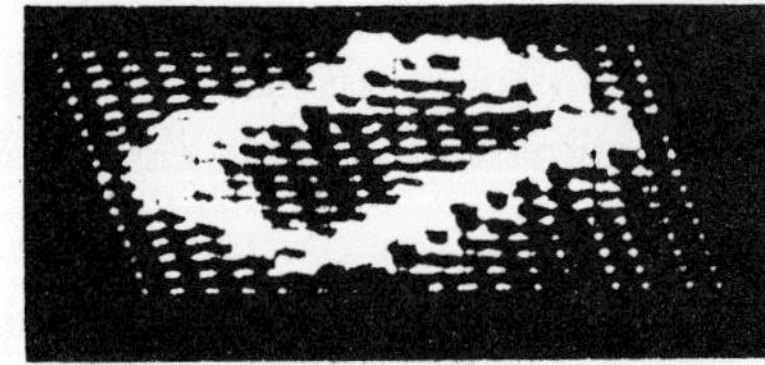

Figure 7.5 Typical tactile sensor output image (after Robertson and Walkden, courtesy SPIE, 1983)

An alternative approach to the construction of a resistive tactile sensor is to measure the resistance change horizontally, as shown in Figure 7.6, the main advantage being the absence of an electrode between the resistive material and the applied force which increases its operational life. However, accommodating both electrode patterns underneath the resistive medium necessarily increases the tactel size and therefore reduces the sensor x–y resolution.

7.3.2 Capacitive tactile sensors

This is not a popular technology for tactile sensors at present but may be developed more in future in view of its potential for measuring higher

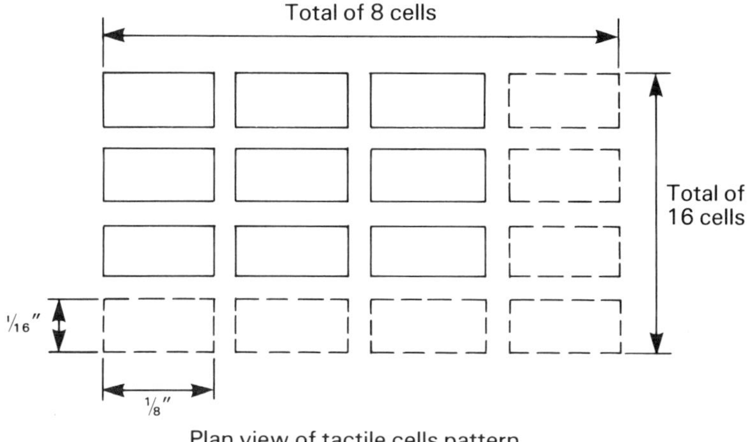

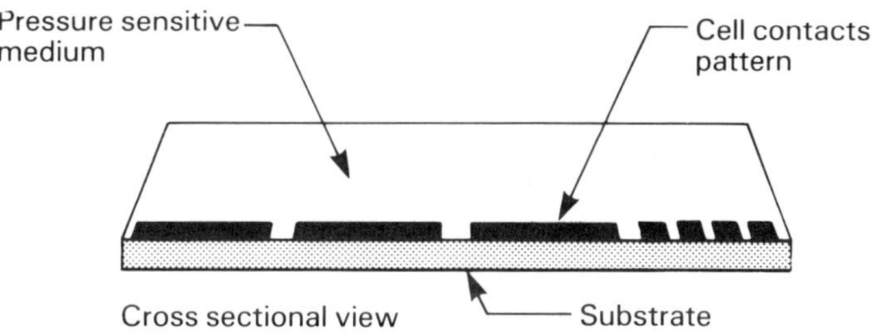

Figure 7.6 Piezoresistive tactile sensor with horizontal electrodes geometry

pressures. This is afforded by the inherently less malleable and therefore more mechanically robust construction of capacitive transducers compared with resistive ones, but at the expense of a bigger tactel size and therefore lower x–y tactel resolution.

The principle of operation of each tactel is predictably the same as that of a capacitive force transducer, as described in Section 4.4, though based on an x–y array of capacitive tactels, as shown in Figure 7.7:

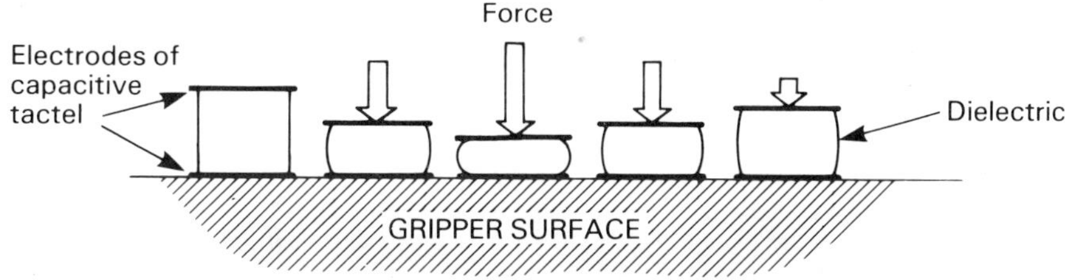

Figure 7.7 Side view of tactile sensor based on capacitive tactels array

7.3.3 Optical tactile sensors

Optical tactile sensors are gaining in popularity as the field of opto-electronics expands from communications and displays to other engineering applications, bringing with it established and therefore easily available, low cost technology. The operation of a tactile sensor based on an optical principle is illustrated in Figure 7.8 and relies on changing the light absorption characteristics of the transparent medium under the effect of an applied force.

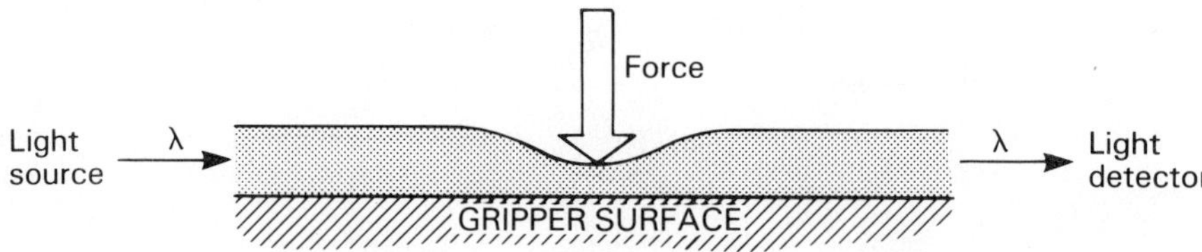

Figure 7.8 Principle of operation of optical tactile sensor

Several clear plastics and glass fibres exhibit this property, so one implementation of an optical tactile sensor is based on an $x-y$ array of light sources and detectors communicating via glass fibres or arranged on the edge of a square sheet of the aforementioned clear plastic, as shown respectively in Figure 7.9(a) and (b). Another optical technique which can be used to build a tactile array is shown in Figure 7.10 and is based on the slotted

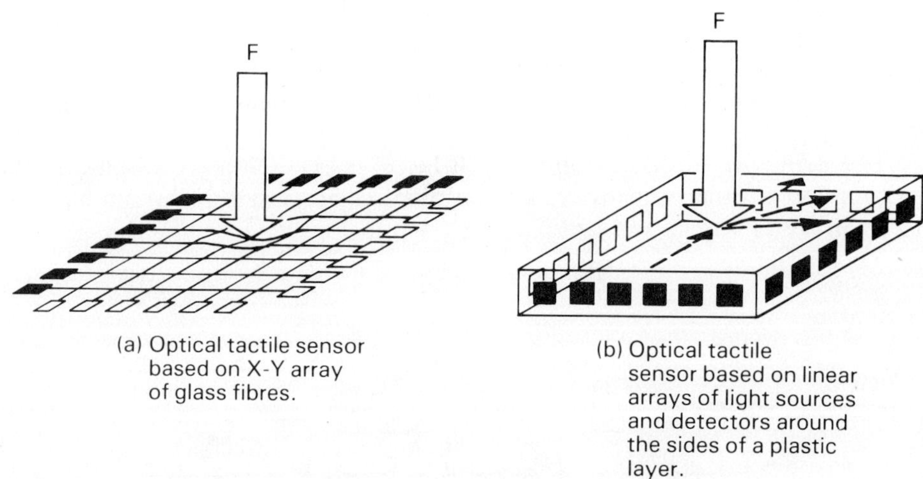

Figure 7.9 Optical tactile sensors operating on light scattering (note: ■ represents a light source and □ represents a detector)

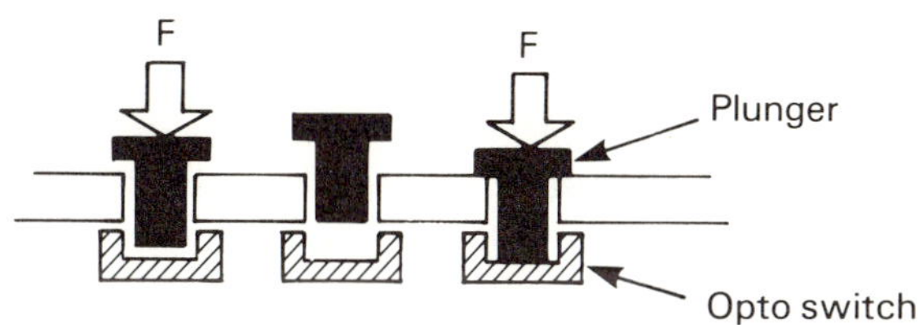

Figure 7.10 Optical tactile sensor operating on light absorbtion (cross-sectional side view)

optoswitch, where the force is applied to a plunger within each individual tactel and partially interrupts the light path between an LED and a photodiode whose output thus provides a measure of the applied force. A tactile sensor based on this principle is available from the Lord Corporation and its agents worldwide.

The resolution and sensitivity of these optical tactile sensors varies considerably but, on the whole, is lower than those afforded by the resistive types.

However, the ease of operation and the relatively inexpensive associated electronics are major advantages and may provide a good reason for use within a robot multisensory feedback system.

Another technique for optical tactile sensing has been described by researchers at the Massachusetts Institute of Technology (Schneiter and Sheridan, 1984) and provides a much higher tactel resolution at the expense of system complexity and cost. The proposed sensor, in fact, uses optical fibres and a standard television camera to provide a visual image of the mechanical deformations produced on the force sensitive 'skin' which is made up of a clear elastomer coated by a reflective silicon rubber. The optical fibres used are much smaller than the typical tactel based on the opto-switch or the capacitive principle and therefore afford a much higher tactel resolution (2100 touch sensitive 'spots' per square inch have been reported in the literature).

7.4 Interfacing of tactile sensors

These devices are still mostly at the research stage, their interfacing depend largely on the technology used and therefore needs to be referred to the manufacturer or research institution data sheets and device manuals.

The few tactile sensors commercially available, however, are usually fully interfaced to one of the popular bus systems and can therefore be accessed like any computer peripheral. An example of such a tactile sensor is the device manufactured by the GEC-Hirst Research Laboratories (Robertson and Walkden, 1983) which already contains an IEEE Multibus interface.

7.5 Conclusions

Resistive type devices, based on a piezoresistive material, are the most popular and versatile tactile sensor, at present. The lower overall cost and ease of operation of optical tactile sensors, on the other hand, is attracting considerable interest and may in future significantly alter this scenario.

7.6 Revision questions

(a) Discuss the application of a robot to pick eggs from a carton using only vision or tactile sensors (not both at the same time) and a standard gripper configuration, that is not a suction gripper.

(b) Describe the principle of operation of an optical tactile sensor. Can you suggest other suitable designs other than those found in this book?

(c) Analyse the action of picking a cup full of liquid and design, in block diagram form, a suitable multisensory system that would help a robot to carry out the operation reliably at high speed.

7.7 Further reading material

1. Robotics, an introduction, D. McCloy and M. Harris, Open University Press, 1986
2. Robotic technology, A. Pugh (editor), Peter Peregrinus Ltd., 1983
3. Industrial robots: computer interfacing and control, W. E. Snyder, Prentice-Hall, 1985
4. Industrial robots: design, operation and applications, T. W. Polet, Glentop Publ. Ltd., 1986

PART III

IMAGE PROCESSING

Chapter 8

Image processing

8.1 Overview

The previous chapters have dealt solely with the data acquisition aspect of sensory feedback, that is they described how to obtain the sensory data (which will be referred to as the 'image data') and how to interface the transducer or sensor to the robot control computer. This image data must, however, be processed in order to extract the required object parameters and supply them to the robot controller so that the necessary action can be planned, thus closing the sensory feedback loop, as shown in Figure 8.1.

This processing can be carried out by the robot controller or, as is becoming increasingly more common, by some local microprocessor contained within the sensor interface, thereby creating what is known as an 'intelligent' sensor. In general terms this combination of the sensory data acquisition and the subsequent processing is referred to as machine perception (Nevatia, 1982) or, more specifically as in the case of vision sensors, as machine/computer vision (Zeuch and Miller, 1986; Ballard and Brown, 1982).

Typically a robotics vision system would be used for one or more of the following purposes:

Motion Control—e.g. modifying a programmed sequence of robot movements to compensate for measured object position.

Quality Control—e.g. in determining whether a part has been produced to the correct tolerances.

Object Recognition—where one of a set of robot programmes is selected depending on the present part.

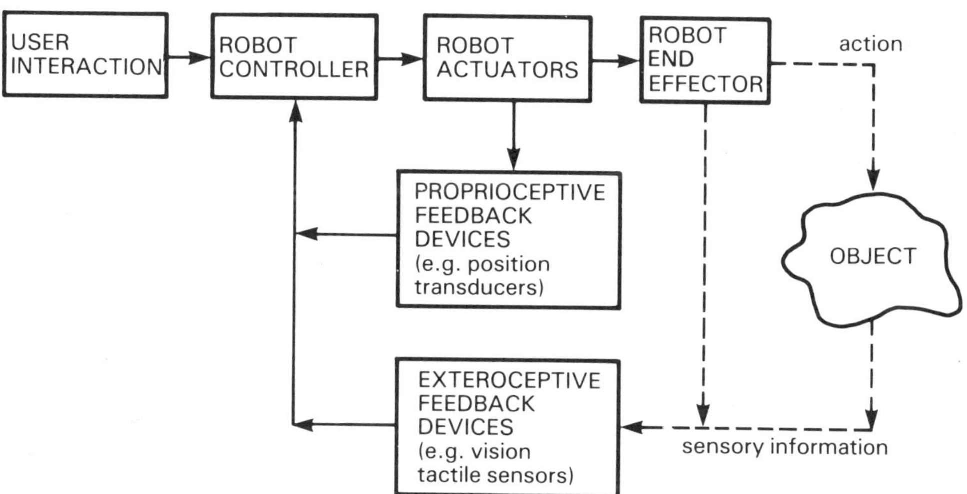

Figure 8.1 Generalised diagram of second generation robot system

To fulfill these objectives a vision system might be required to provide two kinds of output to the other components of the control system:

Measurements—of an object position and orientation, or of part dimensions.
Classification—whether an operation has been successfully performed on a part, or which part has been presented.

In the analysis of image data within the vision system, processes are used which produce both these kinds of output—as classification generally uses measurement data and measurement normally requires the identification of the feature within the image to be measured.

This chapter therefore aims to describe some of the basic techniques used in the processing of the raw image data—that is, it aims to explain the development of the computer algorithms which allow the extraction of the required image parameters, such as object position, size, centroid coordinates and, in the case of 3-D data, the minimum distance from the robot endeffector. The actual choice of the image parameter to be measured, the method and, to some extent, even the hardware required depend on the application. If object width or gap measurement is all that is required, for instance, a linescan camera would be cheaper and easier to use and interface than a 2-D camera; typical examples of such a case (also previously mentioned in Chapter 6) are provided by the arc welding sensor developed by Drews *et al.* (*Drews et al.* 1986; Oldelft, 1985) and the corn cobs inspection plant reported by Hollingum (1984). Other factors which may affect the design of the image processing system might be speed of operation and its flexibility, that is how quickly it can provide the required image parameters and how easily it can be altered to carry out new tasks.

8.1.1 Phases of image processing

The image data provided by the robot transducers and/or sensors, however, is often not in a format suitable for direct image processing and requires the execution of an intermediate pre-processing stage, such as low-pass filtering or the removal of the blur caused by object motion. Image processing can therefore be broken in to two main areas: Image preprocessing and image analysis, as shown diagrammatically in Table 8.1:

Table 8.1 Image processing parsing tree

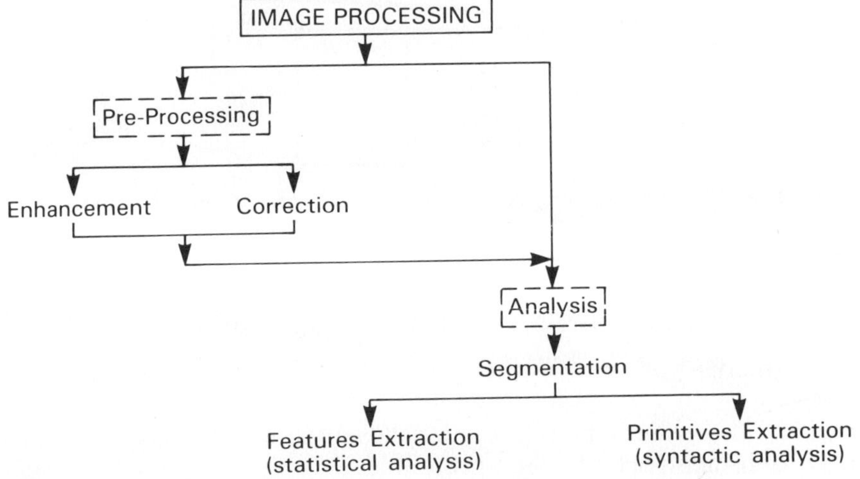

The typical phases of an image processing programme might be:

(*i*) *Pre-processing of the raw image data.*
Necessary to remove the noise generated within the sensor system (such as the electrical noise caused by the A/D conversion) or to counteract the effect of any spatial image degradation (such as the possible geometric image distortion introduced by the vacuum cameras).

(*ii*) *Segmentation of the image into regions of interest.*
These regions would typically correspond to each object, or facet of an object 'viewed' by the sensor.

(*iii*) *Image classification.*
This step usually involves generating suitable descriptions of the image regions corresponding to the object(s) in the sensor field of view.

(*iv*) *Image measurement.*
This is essentially the required end product of most of the image processing carried out in the field of robotics since it is these measurements that enable the required closed loop control in areas ranging from automated inspection to assembly.

8.1.2 A preliminary example of image processing

A simple example of this basic image processing 'routine' is provided by the use of the linescan camera for object width measurement, as shown by Figure 8.2.

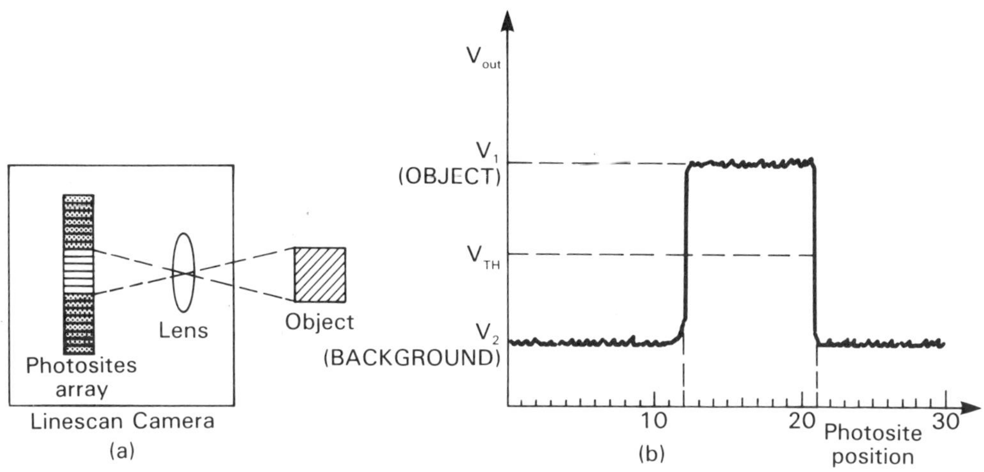

Figure 8.2 Working principle of the linescan camera

The required object width would normally be measured as follows:

(i) *Image pre-processing.* The 1-D image shown in Figure 8.2(b) appears to have a good signal-to-noise ratio. Therefore only a weak low pass filter would be necessary to remove the electrical noise introduced by the shift registers within the linescan camera. These solid state devices are also geometrically very stable so that no image correction is necessary.

(ii) *Image segmentation.* Because of the 1-D simplicity of the linescan camera output the required image region (corresponding to the only object in the field of view) can be separated from the background by a single thresholding operation. This latter could, for instance, make all pixels below V_{th} (a selected output voltage value or 'grey' level) equal to zero. The choice of a suitable threshold value V_{th} is simple in this case (shown in Figure 8.2 half way between V_1 and V_2) but, for more general cases, it is explained in some detail in Section 8.4.

(iii) *Image classification.* Not necessary in this case since there is only one region of interest in this 1-D image.

(iv) *Image measurement.* The required parameter is obtained by counting the number of non-zero pixels. This image measurement can then be related to the actual object width by either a previously carried out calibration procedure or *a priori* knowledge of the sensor optics.

8.1.3 Image degradation

In order to design effective image pre-processing algorithms, some understanding of the image degradation causes (e.g. noise, poorly focused optics, etc.) and their effect on the overall digitised sensor output is required. Image pre-processing, in fact, is essentially an estimation process aimed at producing an 'ideal' image where any degradation has been removed (Figure 8.3) and is based on producing a restored image $r(x, y)$ closely related to the ideal image $f(x, y)$ captured by the sensor. The image degradation can be of two main types:

(a) Point degradation, which affects the intensity value of each image point independently (i.e. it changes the individual pixel grey level). Typical examples of point degradation are electrical noise and transducer nonlinearity.

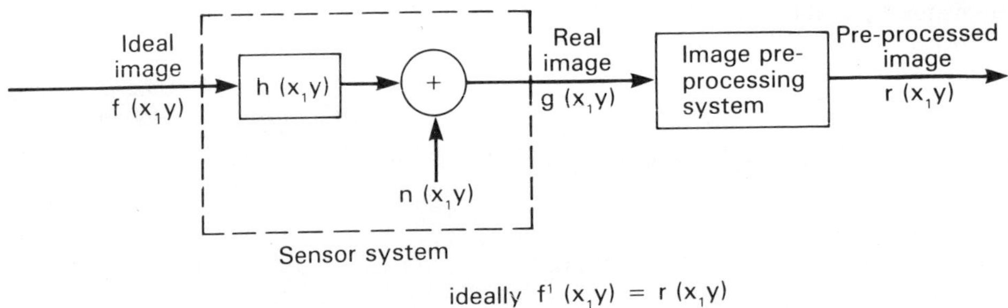

Figure 8.3 Block diagram of typical vision system showing effect of noise and sensor degradation

(b) Spatial degradation, which affects the intensity value of several image points at the same time. Typical causes of spatial degradation are poorly focused optics, object motion (both of which result in blurred images) and geometric distortion (which results in focused but distorted images, such as in the case of pincushion and barrel distortion).

It is worth pointing out that this is not a mere academic distinction: identifying the cause of the image degradation can sometimes provide a prompt solution to the problem without the use of extensive (and therefore expensive) software. For instance if the image degradation is due to spatial distortion caused by improperly aligned optics (e.g. oblique camera viewing angles) then it may be possible to re-design the sensor system to reduce the distortion (i.e. reposition the camera). In general, however, image pre-processing requires the use of considerable computing power. There are two basic techniques that can be employed in this case:

(*i*) *Model the degradation process.*
This requires an understanding of the degradation process and its effect on the image data, and can be gained either from *a priori* knowledge about the sensor (e.g. the use of Vidicon cameras leads to non-linear gamma distortion), or estimated from the image features (e.g. poor light reflections from sharp corners leads to a poor signal-to-noise ratio in those areas) or inferred from statistical analysis of the image properties (e.g. a pixels intensity histogram where the peak occurs near the top digital threshold value—255 for an 8 bit grey level output—suggests that the sensor output has probably reached saturation level).

(*ii*) *Model the image.*
This requires *a priori* knowledge of the object(s) within the scene being viewed. The technique can become computationally cumbersome (and thus both expensive and slow) unless the objects are simple geometric solids or have few degrees of freedom within the image (e.g. rectangular objects in industry or stars in astronomy). In spite of its drawbacks, however, this technique could gain wide acceptance within the image processing community because of its similarity to the increasingly popular 'model based' object recognition technique (Silma, 1984); by sharing the object models data base and some of the subroutines the two techniques could, in fact, reduce the required memory size and overall execution times considerably.

8.2 Convolution and filtering

Most of the image pre-processing techniques are based on digital filtering, that is on the removal of some unwanted effect from the 'real' image supplied by the sensor. This filtering is commonly achieved by the use of task-specific integer matrices known as *convolution masks*. This is an important aspect of image processing and is therefore worth a more detailed description.

Convolution is a mathematical operation on two functions of the same variables, for instance $f(a, b)$ and $h(a, b)$. In image processing terms this allows us to combine two functions (e.g. the degraded image function and the filter function) in order to obtain a wanted effect (e.g. a filtered image). This process is carried out by solving the convolution integral shown in eqn (8.1) which is often quoted as simply $(f * h)$

$$g(x, y) = \int_{-\infty}^{\infty} \int_{-\infty}^{\infty} f(a, b) \cdot h(x - a, y - b) \cdot da \cdot db. \tag{8.1}$$

The complexity of this mathematical notation belies the conceptual simplicity of convolution as a filtering technique. This will become clearer when examples of the use of convolution masks are shown in Section 8.3.2 on image enhancement. A useful pictorial view of convolution has been

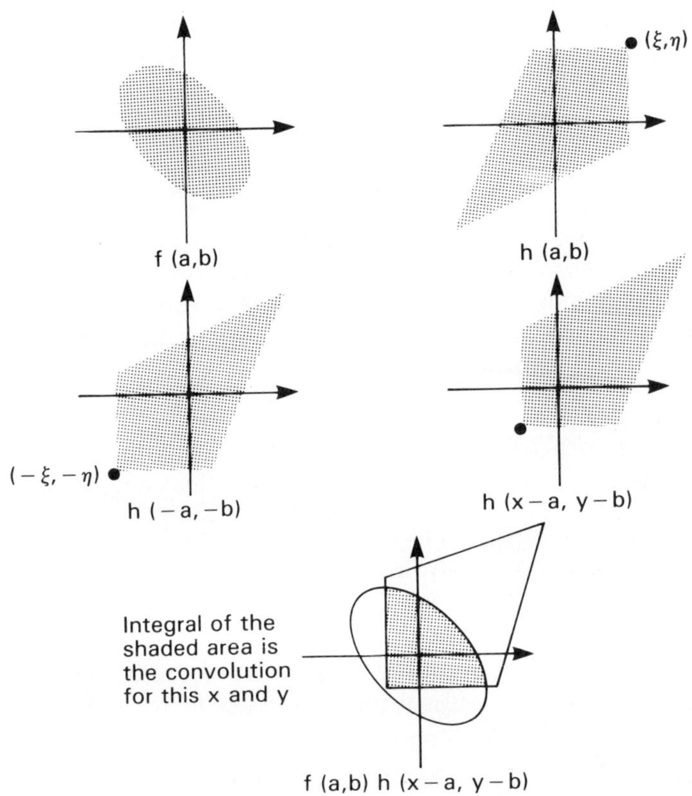

Figure 8.4 Convolution
(After Rosenfeld and Kak, 1982. Courtesy of Prentice-Hall)

proposed by Rosenfeld and Kak and is shown in Figure 8.4. The convolution of $f(a, b)$ and $h(a, b)$, in fact, was shown by eqn (8.1) to be equal to the product of the two functions $f(a, b)$ and $h(a, b)$ with this latter rotated by 180° and shifted by an amount x and y along the two respective orthogonal axis.

Moreover if we consider an image as a function varying in two dimensions, then by analogy with time varying signal analysis, we can perform a harmonic or Fourier analysis on the image. The resultant Fourier transform of the image will give information on the spatial frequency components from which the image is composed.

It is common for some forms of image degradation, for example noise, to occupy a band of spatial frequencies distinct from that occupied by the features of interest. In this case the degradation can be simply removed by transforming the image to the spatial frequency domain, deleting those regions of the transformed image occupied by the degradation component frequencies by multiplying with a suitable mask function, and transforming the result back to the spatial domain.

However this is a computationally expensive process and can be greatly simplified by the use of a suitable convolution. This is possible because there is a simple relation between convolution in the space domain and multiplication in the spatial frequency domain. Thus for two functions f and h

$$f * h = \mathscr{F}^{-1}(F, H) \tag{8.2}$$

where

$$F(u, v) = \mathscr{F}(f(x, y))$$
$$H(u, v) = \mathscr{F}(h(x, y))$$
$$\mathscr{F}(g) \text{ is the Fourier Transform of } g,$$

that is convolving an image function f with a function h is equivalent to Fourier filtering f in the frequency domain with H.

To illustrate this point let us use the same technique to filter the output of a 2-D camera. This output image (sometimes also called the 'observed' or 'real' image) can be represented by the convolution operation, as shown by eqn (8.3), which shows the effect of some image degradation function $h_d(x, y)$ and electrical noise $n(x, y)$ (Figure 8.4).

$$g(x, y) = \int_{-\infty}^{\infty} \int_{-\infty}^{\infty} f(a, b) \cdot h_d(x - a, y - b) \cdot da \cdot db + n(x, y). \tag{8.3}$$

The required filtered or 'reconstructed' image $r(x, y)$ is obtained using convolution (Pratt, 1978) as shown in eqn (8.4)

$$r(x, y) = [f(x, y) * h_d(x, y) + n(x, y)] * h_r(x, y), \tag{8.4}$$

where $h_{r(x,y)}$ is the restoration filter and the symbol '*' indicates 'the convolution of' the two functions that it links. This is illustrated by Figure 8.5.

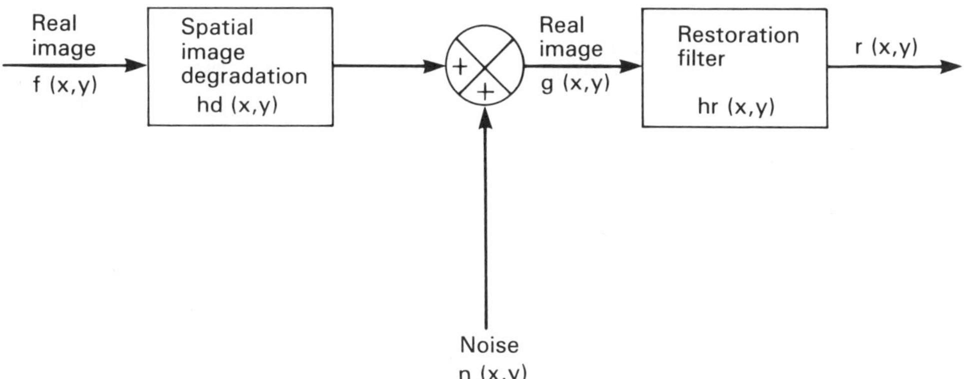

Figure 8.5 Image correction principle

In order to design this restoration filter we need to solve eqn (8.4) for which we can use the F.T. as previously described and as shown in eqs (8.5) and (8.6)

$$\mathscr{F}[r(x, y)] = \{\mathscr{F}[f(x, y)] \cdot \mathscr{F}[h_d(x, y)] + \mathscr{F}[n(x, y)]\} \cdot \mathscr{F}[h_r(x, y)] \quad (8.5)$$

$$R(u, v) = [F(u, v) \cdot H_d(u, v) + N(u, v)] \cdot H_r(u, v). \quad (8.6)$$

This leads to the derivation of the wanted restoration filter $H_r(u, v)$ in the spatial frequency domain. This technique is used, for instance, to obtain the *Inverse Filter,* that is a function which aims to remove any image degradation in order to produce a restored image $r(x, y) = f(x, y)$. This technique is therefore a form of image correction as will be described later in Section 8.3.1.

8.3 Image pre-processing

Image processing was shown in Table 8.1 to be composed of two main sections: image pre-processing and image analysis. The latter will be treated in detail in Section 8.4 whereas image pre-processing is summarised in Table 8.2 from which it can be seen that the two main areas of image pre-processing are image correction and image enhancement.

8.3.1 Image correction

Image correction is required whenever image degradation occurs within the sensor system that is as shown previously in Figures 8.3 and 8.4 whenever the sensor output image $g(x, y)$ is not the same as the ideal image $f(x, y)$. Image correction is therefore based on obtaining a 'restored' image $r(x, y)$ such that $r(x, y) = f(x, y)$. In order to obtain this restored image $r(x, y)$ one needs to process the degraded version $g(x, y)$ with a suitable function $h_r(x, y)$ so as to remove the effects of noise $n(x, y)$ and any other degradation effect. The simplest technique for obtaining the required image restoration is to use an inverse filter, that is a filter whose F.T. is the inverse of the F.T. of the degradation transfer function, as shown in eqn (8.7):

$$H_r(u, v) = \frac{1}{H_d(u, v)}. \quad (8.7)$$

The spatial frequency spectrum of the restored image is therefore given by eqn (8.8) obtained by substituting eqn (8.7) back into eqn (8.6):

$$R(u, v) = F(u, v) + \frac{N(u, v)}{H_d(u, v)}. \quad (8.8)$$

This appears to be a simple yet effective image correction technique since it generates the frequency spectrum of the restored image $R(u, v)$ with an

Table 8.2 Typical breakdown of image pre-processing techniques

IMAGE PRE-PROCESSING

CORRECTION

ENHANCEMENT

MODELLING THE IMAGE

MODELLING THE DEGRADATION PROCESS

HYSTOGRAM OPTIMISATION

MAPPING OPTIMISATION

POINTS DEGRADATION CORRECTION

SPATIAL DEGRADATION CORRECTION

IMAGE ALTERATION

FILTERING (CONVOLUTION)

REMAPPING AND GEOMETRIC INTERPOLATION

BACKGROUND SUBTRUCTION

FILTERING (CONVOLUTION)

LINEARISATION (LOOK UP TABLES)

LOW PASS FILTERING

HIGH PASS FILTERING

OTHER IMAGE FILTERS (e.g. MEDIAN FILTER)

apparent reduction in the noise frequency spectrum ($N(u, v)$ in fact is divided by $H_d(u, v)$). The actual restored image in the 2-D space domain can, of course, be obtained by taking the inverse F.T. of eqn (8.8), as shown in eqn (8.9):

$$r(x, y) = f(x, y) + \left[\frac{1}{4\pi^2}\int_{-\infty}^{\infty}\frac{N(u, v)}{H_d(u, v)}\cdot e^{ij(ux, vy)}\cdot du \cdot dy\right] \tag{8.9}$$

In the absence of any source noise (that is with a sensor that only produces geometric degradation and where the noise, if any, is injected at a later stage) the terms in the square brackets become negligible and a good image reconstruction is carried out, (the re-constructed image $r(x, y)$ is close to the ideal image $f(x, y)$). In the presence of source noise, however, a considerable additive error can be generated by this filtering technique, which would have the practical effect of a poor performance at reconstituting for spatial degradation in the high detail regions of the image.

Another problem with the inverse filter is that if the transfer function of the degradation function $h_d(x, y)$ has zero valued frequency components then the inverse filter cannot be implemented. This can be shown intuitively by referring back to eqn (8.7) noting that for frequency values where $H_d(u, v)$ is zero the relative frequency component for the inverse filter $H_r(u, v)$ needs to have infinite magnitude which is, of course, impossible to obtain.

Other, though more complex, convolution filters for image correction are based on the Weiner Deconvolution Theorem (though this is only suitable for the removal of linear, position independent, blurring functions) or the Coordinate Transformation Restoration (which can cope with position dependent blurring functions such as those caused by rotary motion). A full treatment of such image filtering techniques is beyond the scope of this book but suitable further reading material is listed at the end of the chapter.

A different approach is required in the case of the spatial degradation caused by *Geometric Distortion,* such as those produced by non-linearities in the vacuum camera scanning circuits (e.g. pincushion and barrel distortion) or by wide angle optics (e.g. fish-eye lens distortion). Correcting for geometric distortion, as shown in Figure 8.6, is essentially an image

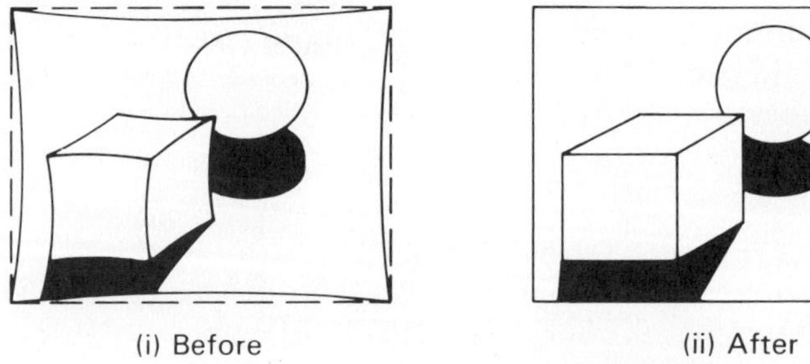

Figure 8.6 Geometric distortion

re-mapping operation aimed at counteracting the image degradation process (i.e. re-position the 'distorted' pixels back where they should be).

The operation is normally carried out in three stages:

(i) the distortion is analysed to obtain the spatial transformation algorithm that relates the 'movement' of each pixel,
(ii) the algorithm is applied to the distorted image and the corresponding output is re-mapped on the same image matrix,
(iii) the grey level values of the re-mapped image are checked to ensure that they lie at integer values on the x–y image matrix; if not, as in most cases, they need to be interpolated.

In order to complete steps (i) and (ii) successfully (i.e. to obtain the most accurate re-mapped image in the shortest possible time) the spatial transformation needs to preserve the connectivity within the image (i.e. it must not allow lines to break up).

Geometric correction is theoretically useful but computationally very expensive. Consequently, in robotics applications it is rarely used, as compensation for distortion is computed at the level of the final measurements instead.

8.3.2 Image enhancement

This aspect of image pre-processing is aimed at transforming the image so to make it more suitable for further processing, either by computer or by human means. Common examples of image enhancement are found in medical imaging (e.g. contrast enhancement) and meteorological satellites imaging (e.g. pseudo colour mapping).

There are three main types of image enhancement techniques in thc general field of machine perception:

(a) Grey-scale modification,
(b) Image transformation or filtering,
(c) Multi-image techniques.

However, grey-scale modification does not add information to the image and is predominantly used to improve the visual appeal of the image data prior to human interpretation. Thus technique is therefore seldom used in the robotics field except for applications where a human operator is involved in a supervisory capacity such as in the inspection of dangerous sites using mobile robots.

8.3.2.1 Grey-scale modification

This technique remaps the grey scale values of pixels to optimise the presentation to a human observer, that is each original grey level is assigned, a new value. Clearly this operation is independent of the value of neighbouring pixels.

The operation may be performed to compensate for non-linearities in the sensor, or to match a sensors sensitivity scale to a human scale, as in the log transform where

$$I_{out} = \log (I_{in}). \tag{8.10}$$

An extension of this approach is the pseudo-colour mapping commonly used in the presentation of land survey satellite images.

The mapping may be derived from the statistics of the image histogram as, in general, the human eye finds the elements of an image most clearly distinguishable when the image has been modified to give a uniformly level histogram expanded to occupy the full available grey scale. This and related techniques of contrast enhancement are commonly called histogram modification.

8.3.2.2 Image transformation/filtering

This form of image enhancement relies on the use of computational (or 'digital') filters to change the characteristics of the sensor's output image.

It therefore shares some of its techniques with the image correction area. The most commonly used techniques of image alteration are low pass filtering and high pass filtering.

(i) *Low pass filtering*—this is an example of 'neighbourhood operation' and is based on calculating the intensity level of the wanted pixel from its original value plus the addition of selected amounts from its neighbours. This technique is used extensively to smooth out any unwanted rapid change in image intensity and therefore also acts as a form of *Noise Reduction* technique. It closely resembles mathematical integration and tends to produce slightly blurred images.

The technique relies on taking an average of the pixels contained within a chosen matrix (or 'mask') with appropriate weightings and placing the result of the averaging operation in the processed image at the same coordinates as the centre of the mask; by moving the mask along the whole image can thus be filtered and a 'new' processed image (with a slightly smaller perimeter) is created, as shown in Figure 8.7:

Image filters are therefore based on convolution, since they use the product of two input functions (an image region function and a filter matrix) to obtain the required output function (a filtered image region), and because of this analogy image filter matrices are often referred to as 'convolution masks'.

If image blurring needs to be avoided (e.g. to preserve the object edges) then alternative low-pass filtering techniques need to be employed, such as the use of a *Median Filter* which replaces the mask centre pixel with the median value of a set of near neighbour pixels. This tends to produce noise-suppressed (i.e. smoothed) images while retaining most of the edges data, as shown in Figure 8.8:

(ii) *High pass filtering*—another example of 'neighbourhood operation'. It is based on a similar technique as the low pass filter but here the convolution

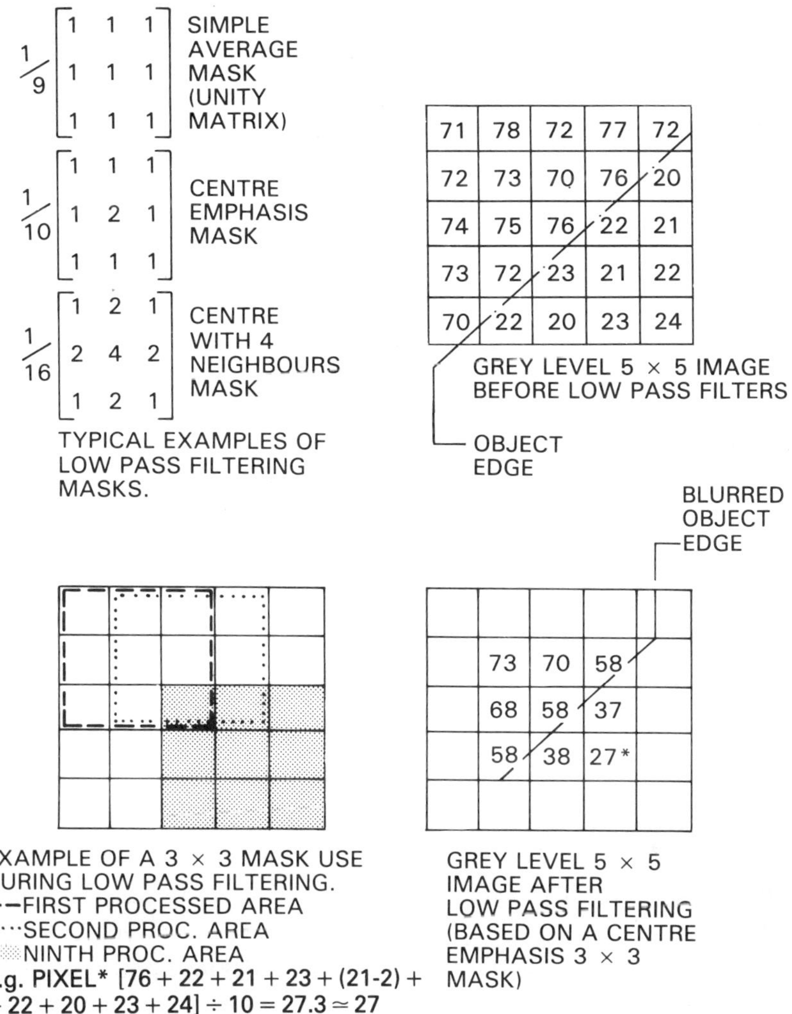

Figure 8.7 Low pass filtering

mask contains negative coefficients. The intensity of the 'filtered' pixel (i.e. the one at the centre of the mask) is therefore obtained by *subtracting* the contributions of the pixel's neighbours from the original pixel at the centre of the mask, as shown in Figure 8.9. This technique is used to highlight rapid changes in image intensities such as edges and contours, it closely resembles mathematical differentiation and tends to sharpen any variation of contrast.

The convolution mask size most commonly used is the 3×3 matrix, since this produces the largest image improvement with the lowest computational load. Larger masks, such as 5×5 or 9×9, are sometimes used in the larger systems but do impose a prohibitive load on the image processing system (typically a 512×512 8-bit grey level image filtered with a 9×9

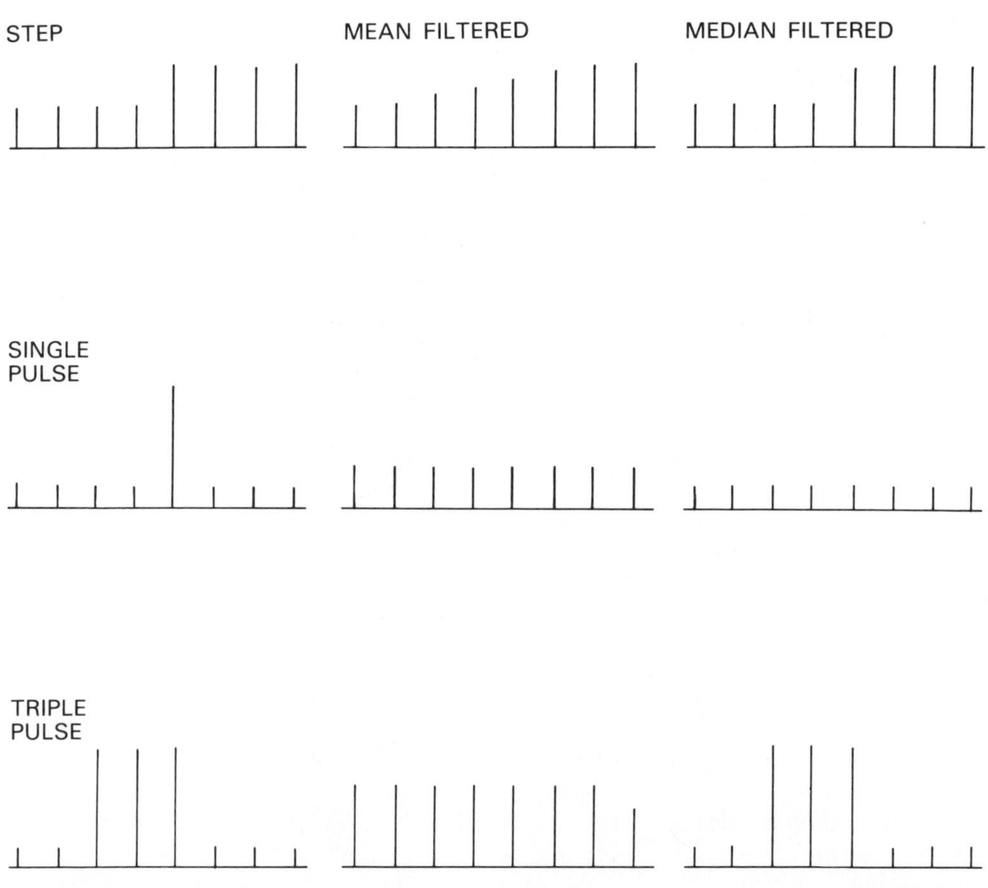

Figure 8.8 Median filtering (compared with the standard low pass averaging filter)

convolution matrix will result in a computational load of approximately 20,000,000 operations).

It is difficult to reproduce computer screen images in a book with such a quality as to be able to display the effect of these image filtering operation. In order to visualise these effects it is helpful, however, to take a 'side' view of an image by plotting the pixel intensities along a particular row or column, as illustrated in Figure 8.10. This shows the effect of smoothing an edge using a low pass filter (Figure 8.10(b)) and of sharpening it using a high pass filter (Figure 8.10(c)).

8.3.2.3 Multi-image techniques

Multi-image techniques achieve this effect by combining two or more images and performing some simple operation, such as subtraction, between pixels in corresponding physical positions in each of the images.

A multi-image technique which is particularly useful in noise reduction is *Frame Averaging*. It applies specifically to the removal of *random noise*

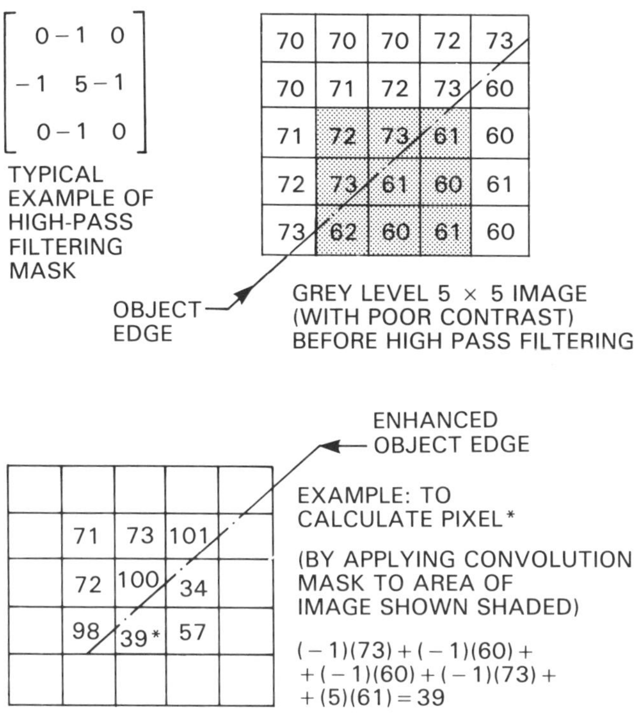

Figure 8.9 High pass filtering

(the kind that produces 'snowy' images even in stationary scenes) which is quite common within electronic systems. The technique is based on averaging the intensity values of corresponding pixels $p(x, y)$ within a sequence of M images; the averaged image $\bar{g}(x, y)$ will thus be given by:

$$\bar{g}(x, y) = \frac{1}{M}\sum_{1}^{M} g(x, y), \tag{8.11}$$

with a corresponding improvement in the signal-to-noise ratio of $\sqrt{M}$.

Another useful technique is background subtraction. Here the image background (i.e. the image without the wanted object(s) in it) is 'subtracted' from the complete image by the use of an algebraic operation (other than addition, which provides averaging). The choice of whether to use subtraction, multiplication or division depends on the specific case. Algebraic subtraction is a simple and quick way to remove unwanted signals, such as lens dirt or defective sensor cells outputs, but division &/+ multiplication may produce better results when the interference also badly affects the object of interest within the image (e.g. when the interference partially covers the object). For binary images these algebraic manipulations can be substituted by the faster logic equivalents such as Exclusive OR for Subtraction, AND for Multiplication and NAND for Division. As well as removing unwanted image irregularities (such as dirt on the sensor lens) this

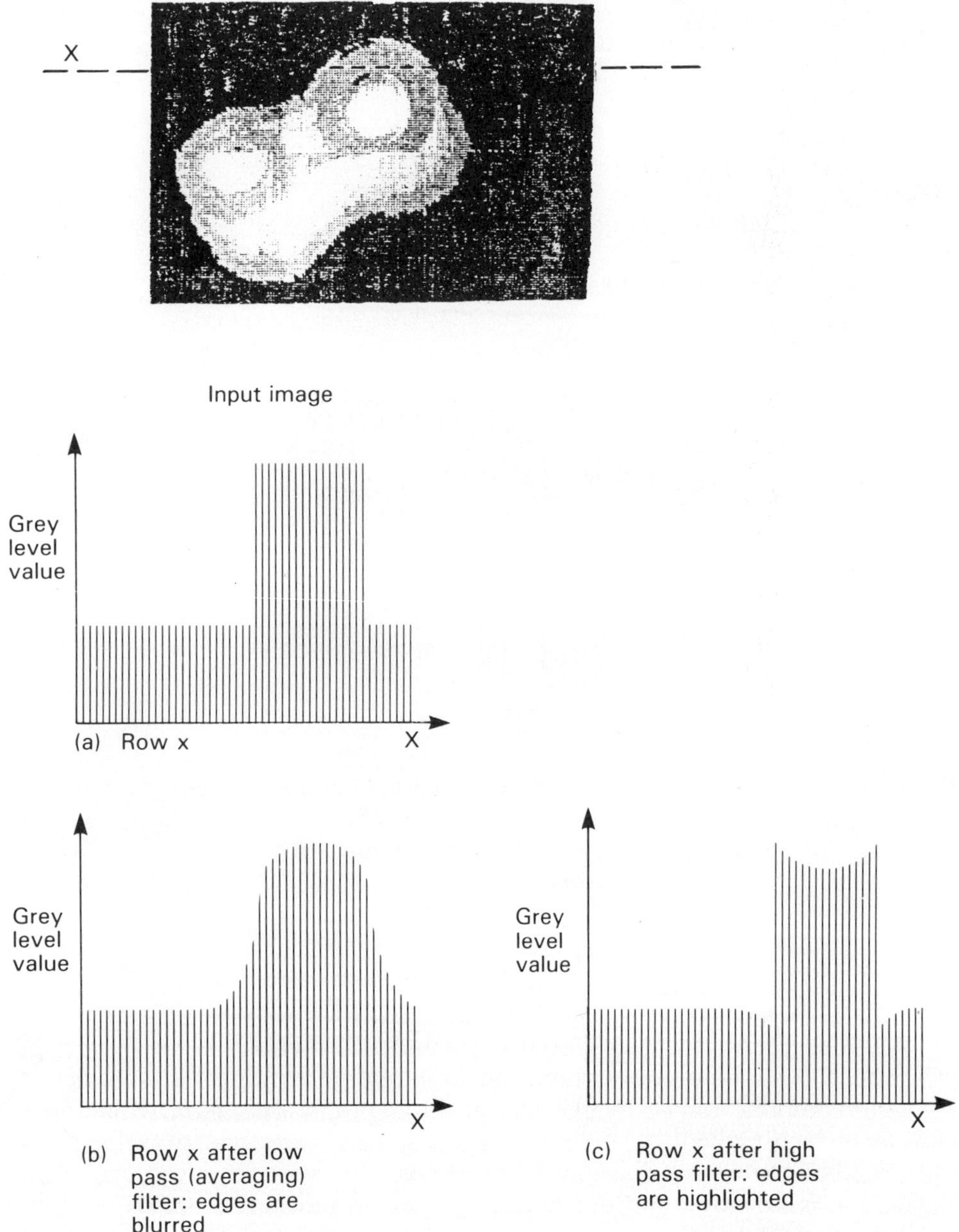

Figure 8.10 Visualisation of filtering effects

technique can help to correct for some sensor non-linearities such as the spatial degradation found in the output image of some sensor whose sensitivity in the centre of the image matrix is different from that of the outer part thus causing what is known as the 'halo' effect, as shown in Figure 8.11:

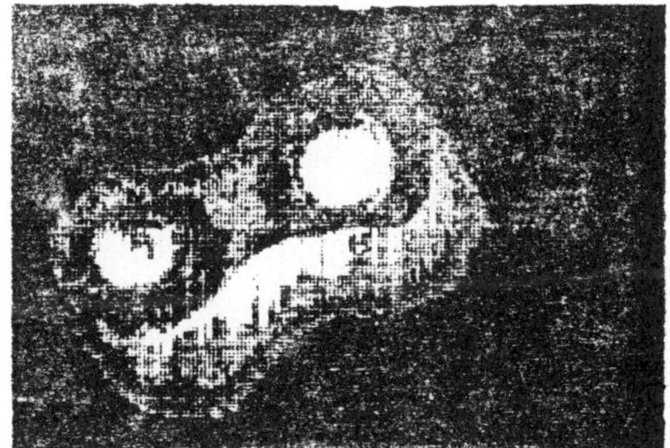

Image corrupted by lens dirt

Background (ie. image with no object)

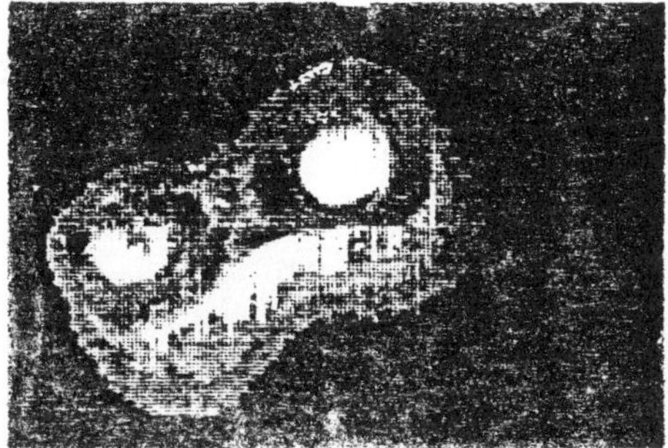

Filtered image

Figure 8.11 Background subtraction

8.4 Image analysis

The end product of the pre-processing techniques thus described is to yield an image which is closely related to the environment sampled by the sensor (i.e. the image correction/enhancement techniques used have, ideally, fully counteracted any noise or degradation processes present within the sensor structure) and which is suitable for further image processing, such as object measurements and features extraction. This area of 'further' image processing is generally referred to as image analysis.

Image analysis is, in fact, used to produce the object characteristics and measurements used by the robot controller to maintain closed sensory loop stability. In other words the object measurements produced by image analysis are used in all areas of robotic applications, from assembly or inspection, to provide the robot controller with the data it requires to make simple decisions and therefore react to any change in object conditions, thus forming the basis for a second generations robot system (PERA reports no. 361 and 366, 1980–81). This may, sometimes, also involve object recognition as in applications where the robot arm has to identify and manipulate a component among several others in an assembly operation. In most cases, however, the production schedule can be arranged so to present the robot arm with only one type of component at a known time thereby avoiding the need for object recognition.

Image analysis was shown by Table 8.1 to be composed of two main parts: image segmentation and image classification (sometimes also called 'image description'). This means that in order to analyse an image one first needs to isolate the required image regions (which usually correspond to the objects within the sensor field of view), then extract the required image parameters from these regions and relate them to the required object measurements. Typical image measurements needed in robot vision applications are: AREA, PERIMETER, SHAPE, MOMENTS OF INERTIA, CENTROID and X-Y COORDINATES.

8.4.1 Image segmentation

Image segmentation aims to distinguish regions corresponding to the required objects within the sensor field of view from the rest of the image.

The segmentation process takes an image as input and produces a description of the regions of interest within that image. This implies two stages to the process; the determination of the region and the generation of a suitable description.

Region determination
As shown in Table 8.3, there are currently two main techniques for region determination: point dependent and neighbourhood dependent techniques.

Point dependent techniques aim to find and sort out regions with similar

Table 8.3 Region determination techniques

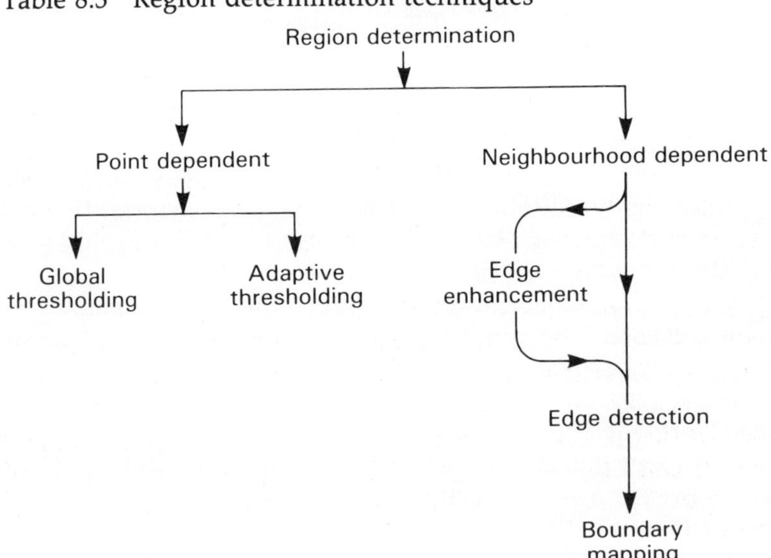

grey level properties (e.g. to separate all 'white' regions from the 'black' ones), this process is commonly known as *thresholding*.

Neighbourhood dependent techniques, on the other hand, aim to find and sort out regions with rapidly changing grey level values (e.g. edges and boundaries), this process is commonly known as *edge mapping*.

8.4.1.1 Thresholding

Thresholding is a point dependent segmentation technique and produces two different types of output images. This tends to lead to a further subdivision in terms of thresholding and semi-thresholding. Given that L_n represents the light intensity of the n_{th} pixel in the image, these segmentation terms can be defined as follows:

Thresholding produces a binary image whose white pixels (logic '1' in optical binary terms) represent inputs above the threshold value T_h. Conversely the black pixels represent inputs below T_h, that is:

THRESHOLDING
(binary output)

$$L_{n(out)}\begin{cases}1 & \text{for} \quad L_{n(in)} \geq T_h \\ 0 & \text{for} \quad L_{n(in)} < T_h\end{cases}.$$

Semi-thresholding produces a grey level image whose pixels intensity retain their input values when they are above the chosen threshold T_h but are replaced by a black level (logic '0') when their value is below T_h, as shown below:

SEMI-THRESHOLDING
(grey level output)

$$L_{n(out)}\begin{cases}L_{n(in)} & \text{for} \quad L_{n(in)} \geq T_h \\ 0 & \text{for} \quad L_{n(in)} < T_h\end{cases}.$$

The selection of the threshold value T_h is quite critical and often requires a human operator to make the choice, through either experience or trial-and-error techniques. This is particularly true in medical image processing where the very complex images encountered require a great deal of skill to reliably extract the wanted image features.

In those fields where a human operator is not part of the sensory feedback loop, as in most systems based on Artificial Intelligence, the computer needs to select the threshold value T_h. This can be done in two different ways:

(i) *Global thresholding,* when the value of T_h is constant throughout the processed image. This technique is usually applied when the wanted object is clearly distinguishable from the background as in the case of images whose grey level histogram is BIMODAL as shown in Figure 8.12. In this instance the choice of threshold value is straightforward since T_h can be placed in the valley between the two peaks produced by the background and object pixel counts, as shown by the shaded area of Figure 8.12.

(ii) *Adaptive thresholding,* when the value of T_h needs to be varied within

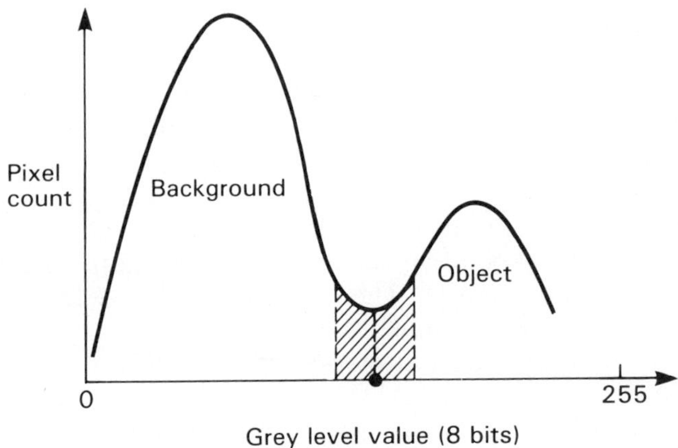

Figure 8.12 Example of bimodal histogram

the processed image. This technique is based on choosing a different value for T_h depending on the grey level background in the particular zone of the image where the wanted object lies. This technique is more common than the global one since most images used in practical applications do not exhibit bimodal histograms. Figure 8.13 shows an example of adaptive thresholding:

In the example shown a single threshold value would not be acceptable because it would produce misleading results; that is either two distorted objects, using only T_{h2} (the shadows would distort the shape of the coin and the chain link would appear bigger than it is), or a single object, using only T_{h1} (the coin would be missing). Adaptive thresholding would allow for the use of T_{h2} while processing pixels in the vicinity of the coin image and the use of T_{h1} when operating on the chain link image, thereby producing a clear and undistorted segmented output.

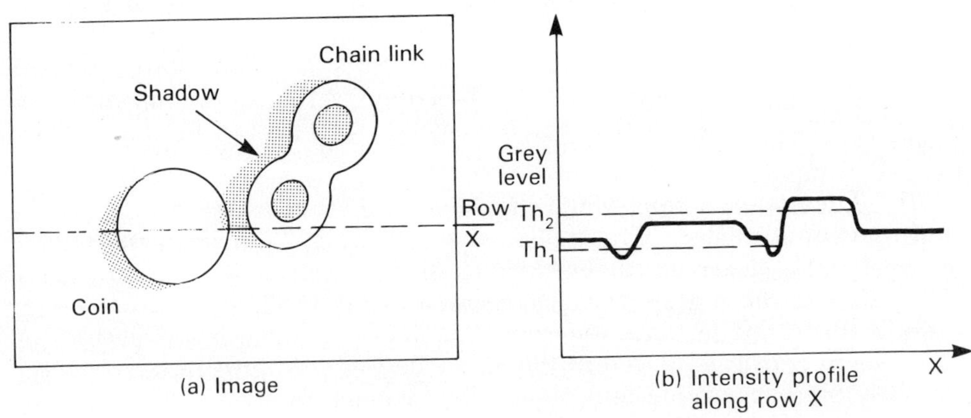

Figure 8.13 Adaptive thresholding example

8.4.1.2 Edge mapping

A more versatile technique, applicable to the wide and complex variety of images encountered in the computer vision field, is edge mapping. This is a neighbourhood dependent technique and, as the name suggests, it is based on processing the image in order to identify the edge pixels and linking them to produce an edge (or boundary) map; that is an image which contains only object boundary information, as shown in Figure 8.14:

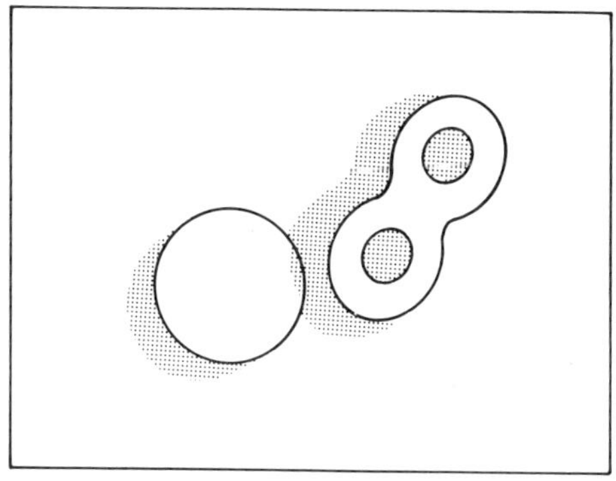

(a) Original grey level image

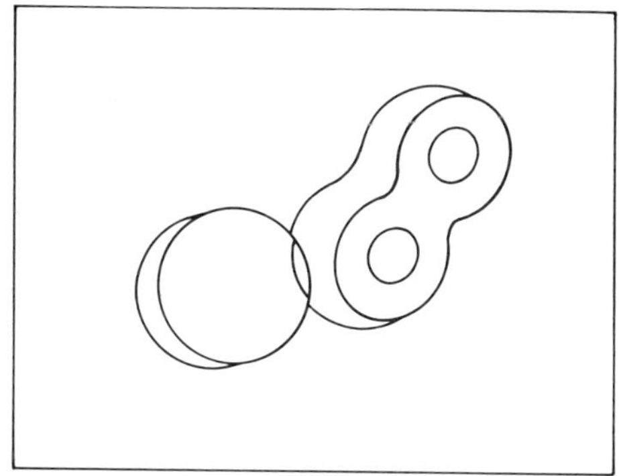

(b) Corresponding edge map

Figure 8.14 Edge mapping example

Edge mapping is essentially a 3 stage operation:

(a) Edges are enhanced by the use of high pass filters; note that this stage may not always be necessary as it may already have been carried out during the image preprocessing option,
(b) Edges are detected by calculating the image first or second derivative,
(c) Object edges are tracked and mapped to produce the 'boundary image'.

The second step, that of edge detection, is crucial to the edge mapping process and is based on mathematical differentiation (that is, on calculating the rate of change of image intensity). It is important to notice that this technique applies to any image and is therefore eminently suitable for range images (i.e. 3-D computer vision) as well as binary and grey level images (i.e. 2-D computer vision).

Edge detection is illustrated in Figure 8.15 by looking at the 1-D image intensity variations along row 'x':

The peaks of the 1st derivative represent the edges of the image along row 'x'. These are identified by thresholding or detecting the zero-crossing point of the 2nd derivative because, mathematically, equating the 2nd derivative to zero is equivalent to finding the maxima and minima of the 1st derivative.

It is worth pointing out that the derivatives of the vertical steps shown in Figure 8.15 are infinite in mathematical terms (since $dy/dx \rightarrow \infty$ as $dx \rightarrow 0$)

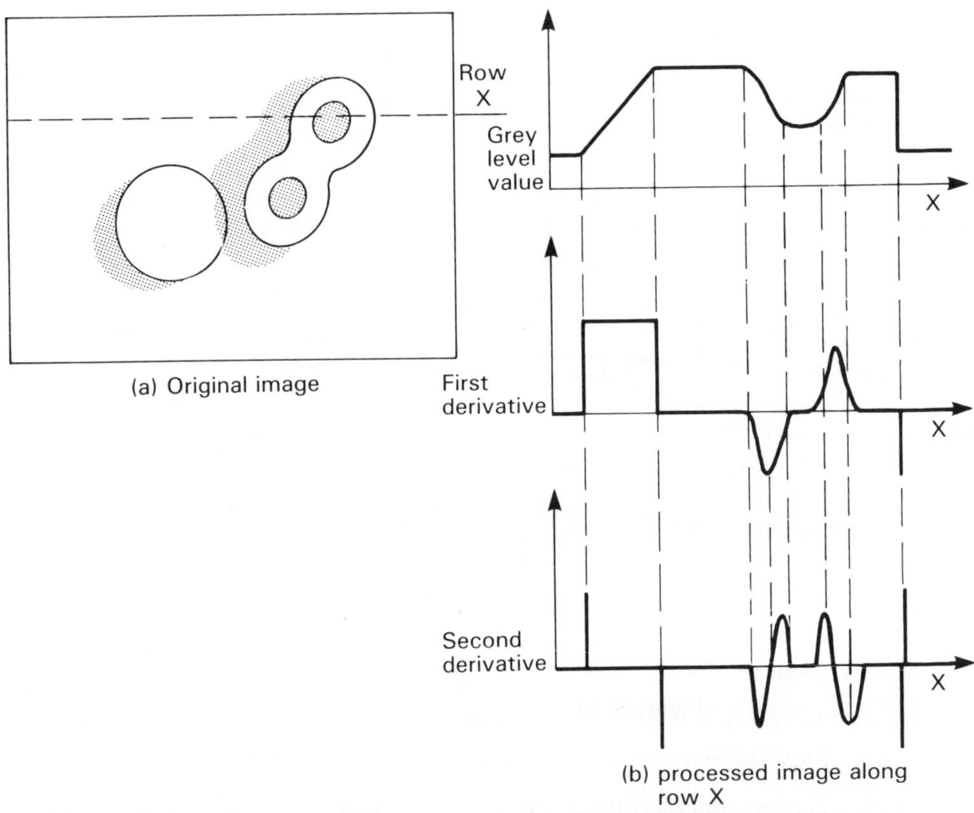

Figure 8.15 Edge detection using 1st and 2nd order derivatives

but are finite in computer terms because in the digital world of computer vision dx is equivalent to the image horizontal resolution which cannot become zero and therefore always produces a finite dy/dx derivative.

One drawback of differentiation is that it tends to highlight any noise present within the image because it treats the noise as a signal and produces an 'edge' pixel wherever a noise spike occurs: this can however be corrected by a limited form of low pass filtering prior to the differentiation stage. The amount of low pass filtering required is small because the edge mapping technique can tolerate a little random noise, since this tends to produce isolated spots which can be eliminated during the next stage in the edge mapping process, namely template matching or boundary tracking.

The edge pixels thus obtained are, in fact, further processed using template matching and boundary tracking masks to check which object they belong to.

Template matching masks are used in the same way as convolution masks and aim to detect a particular configuration of edge pixels. Figure 8.16 shows an example of a template matching mask used to identify edges with a 45° orientation.

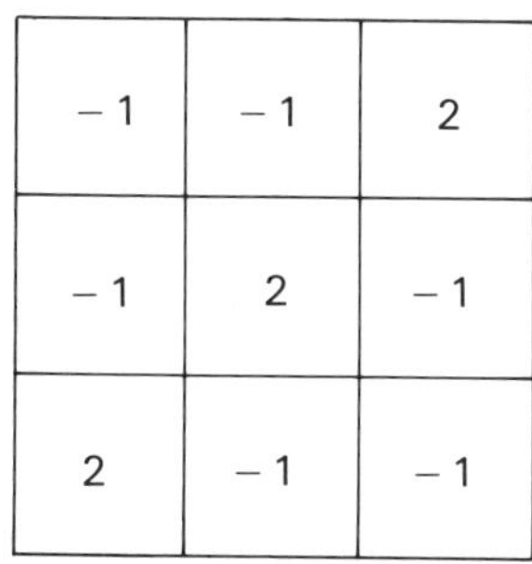

Figure 8.16 Template for 45° diagonal edges

Boundary tracking is applied to the edge data, that is to the image after the differentiation stage, in order to find all the pixels which belong to the same object boundary thus producing the necessary segmented image, as shown in Figure 8.17.

Boundary tracking is therefore based on identifying a pixel on a boundary and on following the boundary line, using a tracking mask, until returning to the same pixel, thus having traced the closed boundary of the object in question. This is achieved in four main steps:

(i) A starting boundary pixel is selected by choosing the highest value in the edge data, that is the highest grey level gradient in the original image,
(ii) A tracking mask, usually a 3×3 matrix, is centred on this starting pixel,
(iii) The next boundary point in the chain is found, as illustrated in Figure 8.18, by examining the pixel diametrically opposite to the current point and its neighbours either side of it,
(iv) The tracking mask is then moved on to this new boundary point and steps (iii) and (iv) are repeated until returning to the starting boundary point as chosen in step (i) thus having traced the object closed boundary.

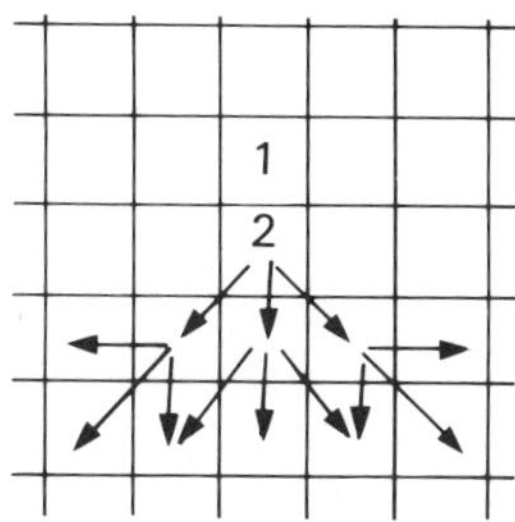

Figure 8.17 Example of boundary tracking algorithm: 1 denotes the starting pixel and 2 the next point found on the boundary

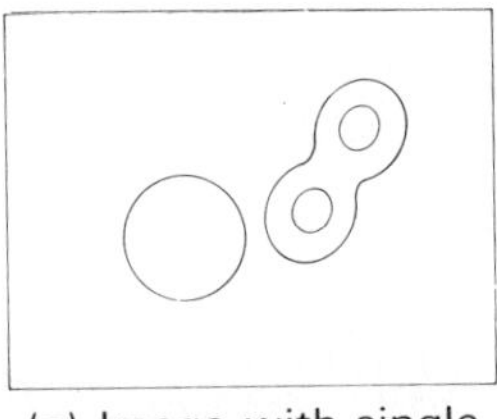
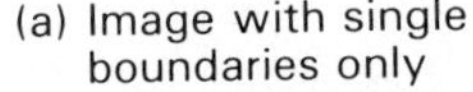
(a) Image with single boundaries only

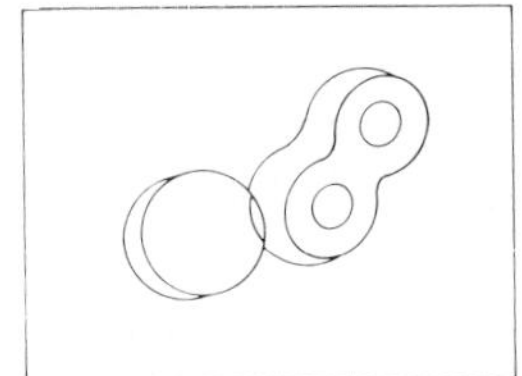
(b) Image with multiple boundaries due to the effect of the shadows

Figure 8.18 The effect of shadows on boundary tracking

If the boundary is single, as shown in Figure 8.18(a), this algorithm will allow the fast tracking of the object edges but if the boundary is more complex (Figure 8.18(b)) as might be caused by overlapping objects, then this algorithm is inadequate and needs some further development in order to avoid the tracking of the incorrect boundary.

Further software development is also required for images whose boundary tracking does not produce a closed curve (e.g. noisy images) since this might produce an endless loop between steps (iii) and (iv) and/or the computer not recognising this broken curve as being part of an object, unless alternatives for a failure to return to exactly the same starting pixels are provided within the software.

8.4.1.3 Generating region descriptions

Given that we have techniques of detecting regions of interest we now need to generate suitable descriptions of these regions. Two kinds of description are commonly used:

Image descriptions—where each region of an image is represented by assigning a distinct value to those pixels lying within the region. So all the regions of interest can be combined in a single image by assigning different 'label' values to the pixels of different regions.

Boundary descriptions—where each region is described by a list of edges between pixels within the region and those outside.

Image descriptions are simply produced from thresholded images by the technique of region labelling.

8.4.1.4 Region labelling

This technique is based on scanning through the image and giving the same unique label to all the pixels belonging to the same region or object. This is achieved by checking that a starting pixel has a grey level value above that of the selected background value and by assigning it a unique label, say 'x'. Subsequent pixels will thereafter be given the same label x only if they are adjacent to at least one pixel with that label.

When determing which pixels are connected either all eight neighbours may be considered (8-way connectivity) or only the 4 horizontal and vertical neighbours (4-way connectivity). Either approach will lead to some contradictory situations, but 8-way connectivity is more commonly adopted. Figure 8.19 illustrates the different results of these two methods.

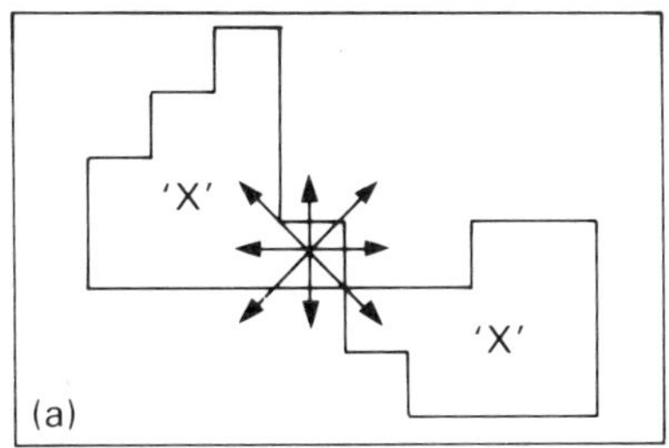

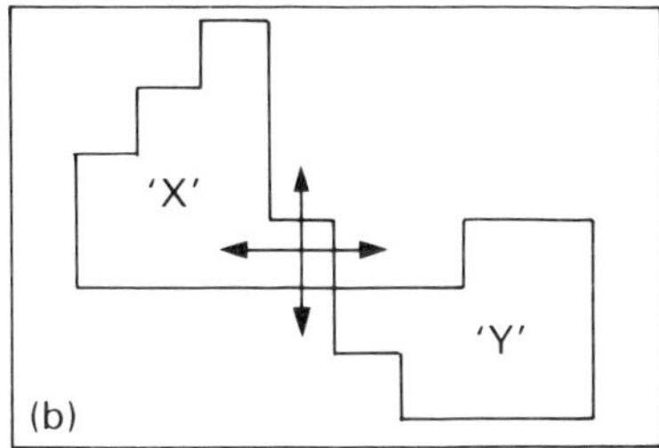

Figure 8.19 Region labelling: connectivity. (a) with 8-way connectivity the two sub-regions are joined (b) with 4-way connectivity the two subregions are not joined

8.4.1.5 Boundary coding

Boundary coding provides a more compact description for a region than image techniques. If, as is common, the boundary is described as the sequence of 'cracks' between a region and its' background then each step can be coded as 'left', 'right' or 'straight', which only requires 2 bits per step.

Clearly it is simple to produce a boundary code description while performing boundary tracking. Furthermore, a very simple tracking algorithm can be used with a thresholded image to produce the boundary code.

8.4.2 Image feature extraction

Once the image has been segmented and the wanted regions within it have been selected, a number of features can be measured. These can then be used for automated object inspection and/or robot closed loop control. A form of object recognition based on the statistical method of object classification can, of course, be carried out if required and is described later in Section 8.4.3.

A wide range of different features may be measured and which are used depends very heavily on the application. For illustrative purposes this section outlines some common measures.

8.4.2.1 Linear and angular measures

Given the coordinates of the points determining the boundaries of measurement, for example a pair of points on either side of an orifice to be measured, it is clearly simple to compute the linear dimension. So the problem in this case becomes one of firstly determining the relevant points and secondly computing this position. These problems can be solved by classification of regions of interest based on simple measurements to be described, and by computing measures for the position of a region.

8.4.2.2 Simple measures

The following measures as a set of simply computed features which can be used for region identification;

(i) *Area and integrated optical density*

The area of a region is simply the number of pixels within the region, normalised by some notion of the area of a pixel.

$$A = k \cdot \sum_{i=1}^{n} 1, \qquad (8.12)$$

where k is the area of 1 pixel and n is the number of pixels. Clearly this is the discrete version of

$$\text{Area} = \iint k \cdot dx \cdot dy. \qquad (8.13)$$

If instead of simply counting pixels we sum their grey levels then we obtain the Integrated Optical Density (IOD).

$$\text{IOD} = \sum_{i=1}^{n} g_i. \qquad (8.14)$$

Again in the continuous case

$$\text{IOD} = \iint f(x, y)\, dx \cdot dy. \qquad (8.15)$$

If we have a boundary coded description of a region we can still compute area from algebraically summing the boundary vectors:

$$\text{Area} = -\sum_{p=1}^{n} \cos \alpha p \cdot yp, \qquad (8.16)$$

where αp is the angle of the pth vector to x axis. For simply coded boundary $\cos p$ can only adopt values of -1, 0 or 1.

The amount of computation included here is considerably reduced as only the boundary vectors are used instead of the complete set of pixels in the image. However note that there is no method of computing the IOD from a boundary description.

(ii) *Perimeter*

The perimeter of a region is simple to compute for either form of region representation. For labeled regions the count of boundary pixels can be taken. This may be the set of region pixels each of which has at least one background neighbour or vice-versa. Neither of these will produce the same measure as that for the boundary description, which is obtained from the arithmetic sum of boundary vector sizes.

Perimeter, although simple to compute, is an unsatisfactory

measure as it is very noise dependent and furthermore increases in an unpredictable way as increased image resolution reveals greater boundary detail. However it finds its uses, particularly in some of the shape measures to be described shortly.

(iii) *Envelopes*
The simplest region enclosing envelope is the rectangle. The minimum enclosing rectangle (MER) is the smallest rectangle enclosing all the region of interest. Typically the best fitting rectangle is computed for a number of orientations and the smallest area rectangle taken. From the MER can be taken a measure of size (MER area) and a shape measure,

$$\text{the aspect ratio} = \frac{\text{Max Side Length}}{\text{Min Side Length}}.$$

Higher order envelopes can be used but are less simple to compute.

(iv) *Derived shape measures*
New measures can obviously be produced by combining those considered so far. Commonly used here are the dimensionless, hence size independent, shape measures:

$$\text{Circularity} = \text{Perimeter}^2/\text{Area}$$

$$\text{Rectangularity} = \text{Area}/\text{MER area}.$$

The circularity measure approaches 4π as the region shape approaches circular. Likewise the rectangularity approaches 1 as the shape approaches a rectangle.

8.4.2.3 Moment based measures

Measurements are also required for the positioning and orientation of regions. A reasonably stable measure of a regions' position is the position of its centroid, calculated by analogy with physical bodies as:

$$\bar{x} = \iint x \cdot f(x, y)\, dx \cdot dy \Big/ \iint f(x, y)\, dx \cdot dy$$

$$\bar{y} = \iint y \cdot f(x, y)\, dx \cdot dy \Big/ \iint f(x, y)\, dx \cdot dy.$$

These measures give rise to the notion of a generalised measure of movement:

$$M_{p,q} = \iint x^p \cdot y^p f(x, y)\, dx \cdot dy$$

$$M_{0,0} = \iint f(x, y)\, dx \cdot dy = \text{IOD}$$

$$M_{1,0} = \iint x \cdot f(x, y)\, dx \cdot dy = \bar{x} \cdot \text{IOD}$$

etc.

For simplicity we can consider binary images produced by labelling such that:

$$f(x, y) = 1$$

for all points in the region of interest, and $= 0$ otherwise. This simplifies the formulae to

$$M_{p,q} = \iint x^p \cdot y^q \, dx \cdot dy,$$

where we perform the integration just over the region of interest, and

$$M_{0,0} = \iint 1 \cdot dx \cdot dy = \text{Area}$$

$$M_{1,0} = \iint x \cdot dx \cdot dy = \bar{x} \cdot \text{Area}$$

$$M_{0,1} = \iint y \cdot dx \cdot dy = \bar{y} \cdot \text{Area}$$

Clearly these properties can be simply evaluated in the case of discrete images. Thus for a labelled, thresholded image

$$M_{1,0} = \sum_{i=1}^{n} x_i$$

$$M_{0,1} = \sum_{i=1}^{n} y_i.$$

Similar formulae can be produced to evaluate these terms from a list of boundary components.

For the purposes of orientation measurement a common measure is the orientation of the principal axis of second central moments. This is the axis through the centroid about which the second moments are minimized and corresponds to what might be described as the best fitting longest axis of the region (see Figure 8.20).

This axis can be computed from the axis aligned second central moments:

$$M_{2,0} = \iint (x - \bar{x})^2 \, dx \cdot dy$$

$$M_{0,2} = \iint (y - \bar{y})^2 \, dx \cdot dy.$$

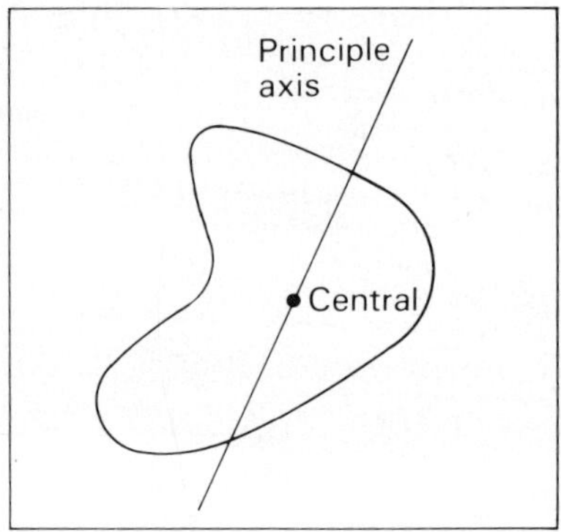

Figure 8.20 Principal axis of 2nd central moments

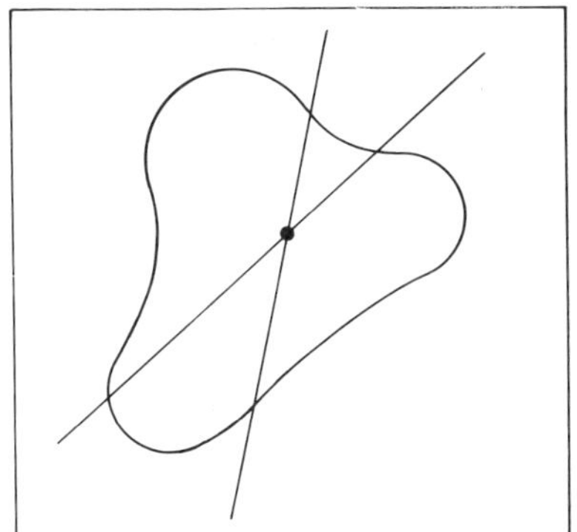

Figure 8.21 Competing principal axes

Some care must be taken in cases where there may be two competing axes between which the result may switch in successive images through simple noise induced variability (see Figure 8.21).

8.4.3 Image classification

The aim of image classification is to deduce a correspondence between regions of an image and some model of the perceived real world to which they correspond. Hence classification is essentially a process of producing a description, though the description may be no more than 'it's a case of *x*'. There are two main techniques of classifying:
(i) statistical classification, and (ii) syntactic classification.

8.4.3.1 Statistical classification

This is by far the most common method currently used in robotics applications. It is based on a statistical model of the measured properties of regions corresponding to different real world objects.

As a simple example consider distinguishing two types of component by area and IOD. The distribution of these types of measures might appear as in Figure 8.22.

This method is simple to evaluate and generally adequate for robotics applications where the objects viewed by the sensor are commonly simple prismatic solids.

However, the output is a simple class case selection with no direct description of any variability in the sub components in the region of interest.

8.4.3.2 Syntactic classification

This approach is based on the generation of a description of a region as composed of a number of sub components, or primitives, together with the spatial relationship of these primitives. For each class of object the acceptable arrangement of primitives is defined by a set of rules—the 'grammar' for that object. See Figures 8.23 and 8.24 for examples.

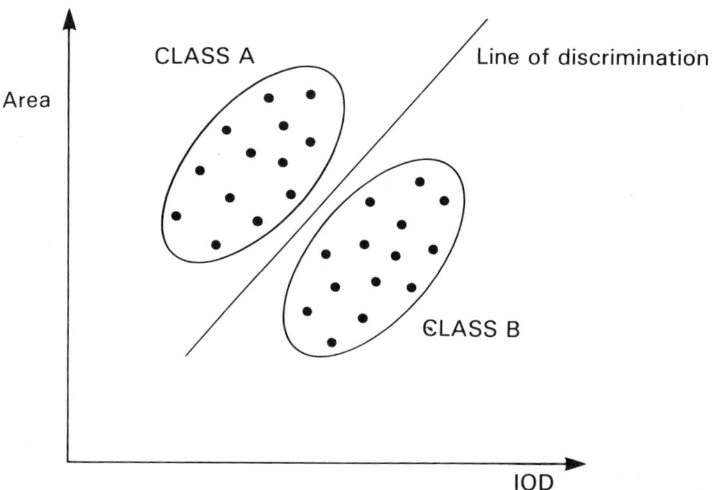

Figure 8.22 A feature-space plot of two features measured for a number of examples of class A and class B objects

The classification process consists of determining which grammar best describes the configuration of primitive components in the image.

This method offers a far richer result as each case classification can be accompanied by a parse tree describing the way in which the region is constructed. Furthermore the grammars allow for a greater flexibility in describing what may be considered as being a member of a given class.

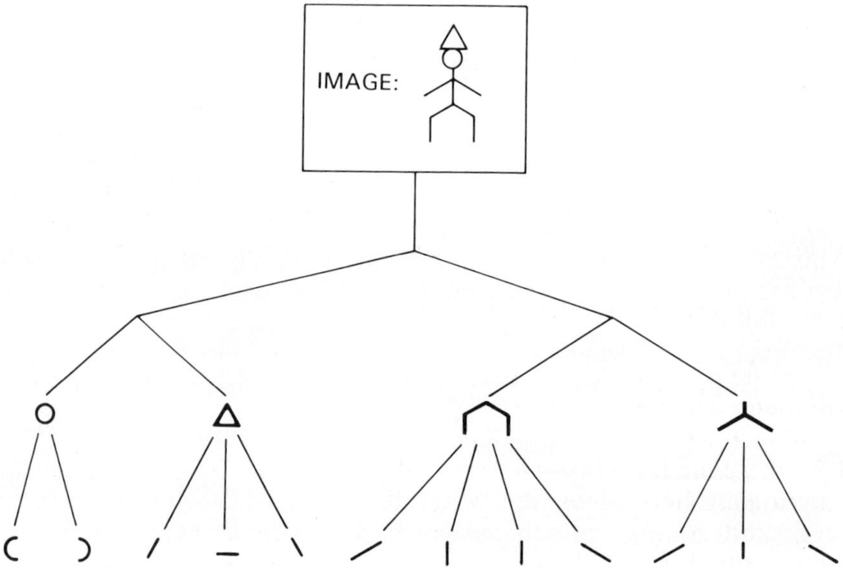

Figure 8.23 Syntactic image breakdown

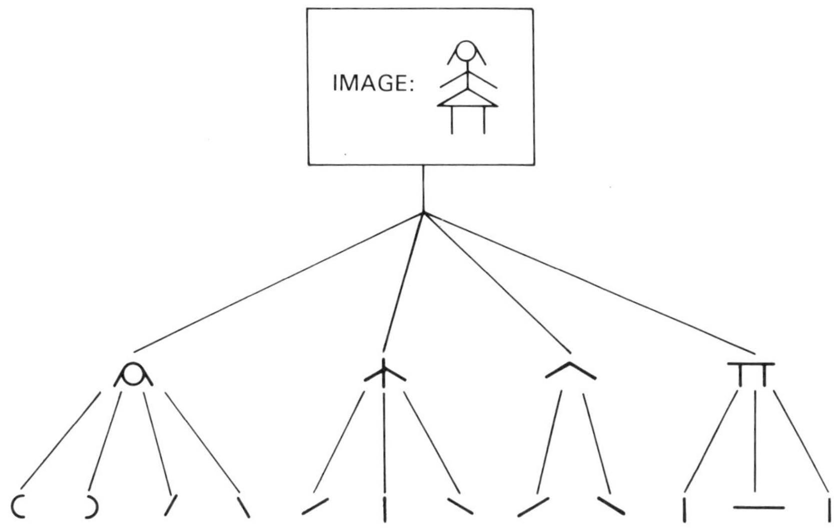

Figure 8.24 Another example of syntactic image breakdown

However, computationally the process is far more complex and it is not simple to produce suitable grammars, which results overall in a less reliable method.

8.5 Conclusions

Image processing consists of two main phases: image pre-processing and image analysis.

Image pre-processing is itself divided into enhancement and correction. Image enhancement is essentially aimed at improving the visual appearance of the image whereas image correction is a technique of compensating for any degradation introduced during the sensing stage and is based mostly on the use of convolution masks to carry out low and high pass filtering operations.

Image analysis is based on two stages: image segmentation and image measurement. Having pre-processed the image to reduce any degradation or noise, in fact, the image then needs to be segmented and the required application dependent features extracted, the most popular being object area, perimeter and centroid.

Derivated object features, such as the circularity and rectangularity shape descriptors, can be used in addition to the primary calculated features to help in object recognition and inspection applications.

8.6 Revision questions

(a) Explain the principle purpose of a high pass convolution mask.

(b) Calculate the area of the object shown on the right using a suitable computer algorithm.

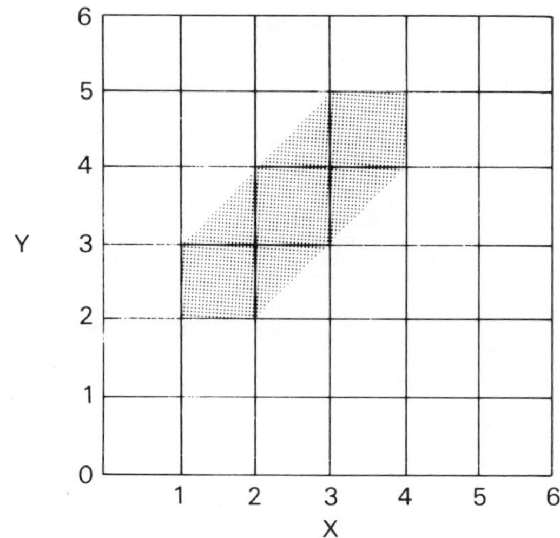

(c) Calculate the output from the image shown in Figure 8.R.c using a low pass convolution filter.

24	25	23	26	85	82
24	21	24	90	88	85
25	25	85	87	86	87
26	84	91	85	83	85
82	89	83	84	85	82
92	87	84	83	81	84

8.7 Further reading material

1. *Digital image processing,* R. C. Gonzalez and P. Wintz, Addison Wesley, 1979
2. *Digital image processing,* K. R. Castleman, Prentice-Hall, 1979
3. *Digital picture processing,* A. Rosenfeld and A. C. Kak, Academic Press, 1982 (Vols 1 & 2)
4. *Digital image processing—a practical primer,* G. A. Baxes, Prentice-Hall, 1984
5. *Image analysis: principles and practice,* Joyce Loebl, 1985
6. *Computer vision,* D. H. Ballard and C. M. Brown, Prentice-Hall, 1982

Solutions to Revision Questions

	Wirewound pot. (typ.)	Optical transd.
angular resol.	< 3 degrees	360/120 = 3 degrees
Oper. life	10,000 rotations	> 10,000 rotations (non-contact meas.)
Cost	< £50 each	> £50 each (typ.)
Size (diameter)	40 mm	40 mm (typ.)

Wirewound are typically cheaper, have the same approximate size and a higher resolution (compared to an optical position transducer with 120 lines) but do suffer from a lower operational life.

2 (b)

Using a radial encoder as shown in Figure 2.8 and applying eqn (2.3) we have:

No. of lines required = 360°/6 = 60°.

We can use a standard photodiode with an active area of 1 mm^2 (e.g. a square diode geometry) but, in order to maintain the 6 degrees resolution we need to mount it at a distance r from the centre, that is:

$$r \geqslant \frac{W_P}{\sin(\Delta\,\alpha/L)} \geqslant \frac{1}{\sin 3^\circ} \geqslant 19.1\,\text{mm} \simeq 20\,\text{mm}.$$

2(c)

The answer is as shown in Figure 2.13 with 5 channels since, using eqn (2.5):

angular resolution $= 360/2^c = 15$ degrees,
therefore

$2^c \geqslant 24$ thus giving $c \geqslant 5$ channels.

3(a)

We need to turn the relay ON when the object breaks the light beam (that is when the cell is in the dark, e.g. for an illumination < 10 lum). Therefore we need to have $V_2 > V_1$ when the cell has a resistance greater than 10 KΩ. That is:

$$\frac{R_3}{R_2 + R_3} < \frac{R_1}{R_1 + R_p};$$

let R_1 be 5 KΩ and therefore $V_2 > 4$ V

$$\frac{12 \cdot R_3}{R_2 + R_3} > 4$$

Suitable values $R_2 = R_3 = 1.5$ KΩ with a 22 KΩ preset resistor used for fine adjustment of the threshold value.

3(b)

Since op. amp. is ideal ignore the input capacitance and resistance. From eqn (3.2) we can extract R_L:

$R_L = 1/(2 \cdot \pi \cdot f_{bw} \cdot C_j) = 318\ \Omega$.

As op. amp. is ideal it has no bias current, therefore:

$V_0 = i_p \cdot R_L = (\eta_{EFF} \cdot \phi_{DIODE}) \cdot R_L$,

which leads to the required optical flux coupled to the photodiode active area:

$$\phi_{DIODE} = \frac{V_0}{\eta_{EFF} \cdot R_L} = 6.29\ \text{mW},$$

coupled into 1 mm^2 or, as is more commonly quoted, 629 mW/cm^2.

4(a)

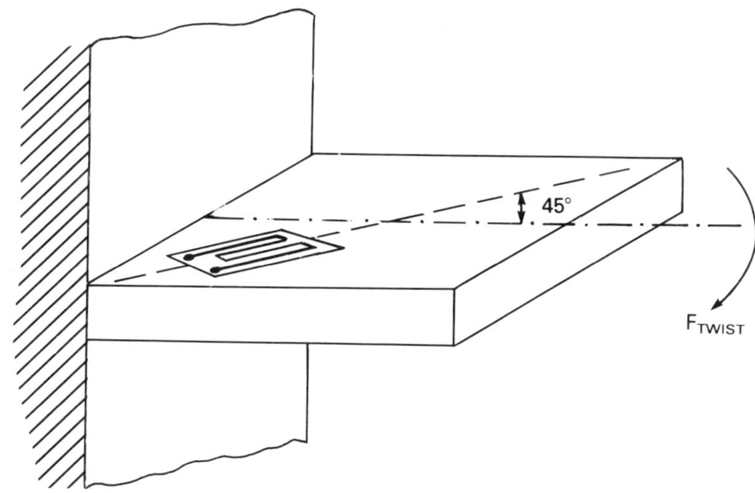

4(b)

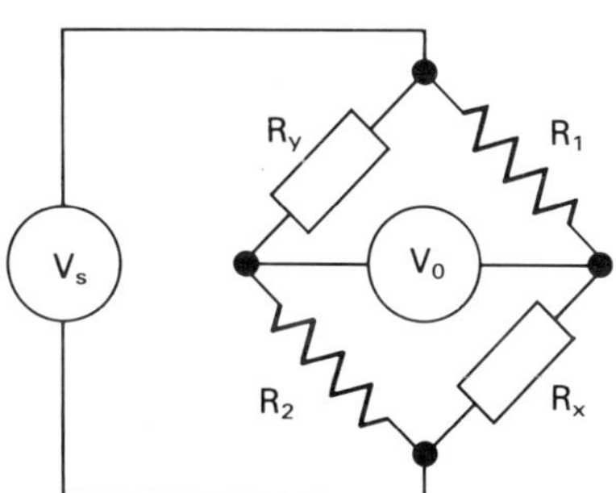

4(c)
From a standard material's data book Young's modulus for steel is:

E_{steel} = Stress/Strain = 2.1×10^5 MN/m,

whereas the nichrome gauge factor can be obtained from Table 4.1:

$G_{f(\mathrm{NiCr})} = 2.0$.

Using eqn (4.12):

$$\Delta V_0 = \frac{V_s \cdot G_f}{2} \cdot \frac{\Delta L}{L} (1 + \mu),$$

and combining it with eqn (4.10) we have:

$$\Delta V_0 = \frac{V_s \cdot G_f}{2} \cdot \left(\frac{\text{STRESS}}{E}\right) (1 + \mu),$$

therefore the answer is:

$$\Delta V_0 = \frac{10 \times 2.0}{2} \left(\frac{0.4}{2.1 \times 10^5}\right) (1 + 0.3) = 24.7\ \mu\text{V}.$$

5(a)
As shown, in Figure 5.1 we have that for $R_L = 1\ \text{K}\Omega$ the tachogenerator output goes from 0 to 10 V, therefore a 1:5 attenuation network is required for interfacing to the A/D converter. A suitable circuit is:

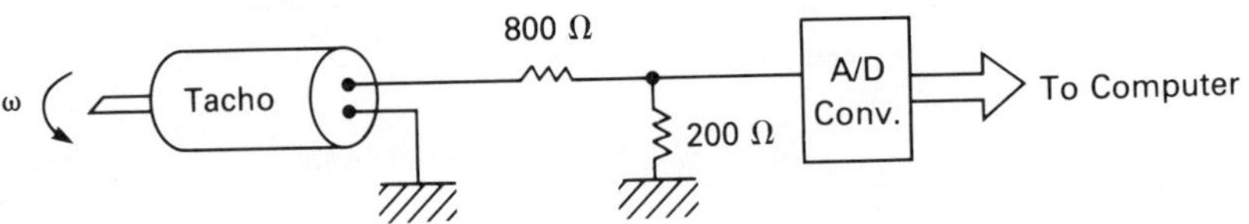

5(b)
A photodiode equivalent circuit is essentially capacitive (see 3.3.1 and 3.3.2) and therefore its output voltage follows a capacitive charge and discharge behaviour:

$V_0 = V_F \cdot (1 - e^{-t/\tau})$,

since 20% of the full output voltage V_F can be assumed to be sufficient for reliable operation we have:

$$\frac{V_0}{V_F} = (1 - e^{-t/\tau}) = 0{\cdot}2;$$

taking logarithms and extracting t:

$t = -(\tau \times \log_e 0{\cdot}8) = 2.23\ \mu\text{sec}.$

This is the minimum transition time from light to dark that the photodiode circuit will tolerate and translates to a maximum frequency of 448 kHz. A disk with 1000 lines on it will therefore achieve this frequency at 448 rev/sec, that is 26880 rev/min.

5(c)
The minimum speed is

$$\text{Min. rev/sec} = \frac{60}{60}\ \text{R.P.M} = 1\ \text{rev/sec}.$$

For a measurement resolution of 10% we need to count 10 pulses per measurement interval. This latter need not be any shorter than 10 msec since the computer could not read it any faster, therefore to measure a minimum speed of 60 rev/min with a resolution of 10% and a measurement interval of 10 msec we need an optical encoder with 1000 lines at least, i.e.:

$$\frac{10\ \text{pulses}}{10 \times 10^{-3}\ \text{sec}} = \frac{x}{1\ \text{sec}} \quad \therefore\ x = 1000.$$

6(a)
Refer to para. 6.2 and, in particular, Table 6.2.

6(b)
Refer to para. 6.5.4(iii). As the speed of light is much greater than that of sound, the time of flight of a light pulse is much smaller than that of a sound wave.

8(b)
Using the zero moments formulae shown in Section 8.4.2.3 and expanding them into discrete form

$$\text{Area} = M_{(0,0)} = \tfrac{1}{2}\sum_{p=1}^{n}(X_p \cdot \Delta Y_p - Y_p \cdot \Delta X_p)$$

$$= \tfrac{1}{2}\,\Sigma\,(-2, 0, 0, 4, 5, 2, 2, -1)$$

$$= \tfrac{1}{2} = 5.$$

8(c)

Using a simple averaging 3 × 3 convolution matrix $\begin{bmatrix}1 & 1 & 1\\ 1 & 1 & 1\\ 1 & 1 & 1\end{bmatrix} \div 9$

we have:

	31	45	66	80	
	45	66	80	86	
	65	79	85	85	
	71	85	84	83	

For example, the top left hand figure is obtained as follows:
(24+25+23+24+21+24+25+25+85)/9 = 30.66 = 31.

References

Agin, G. J. (1972) Representation and description of curved objects (Ph.D. diss.) AIM-173, Stanford AI Lab.

Andre, G. (1985) A multiproximity sensor system for the guidance of a robot end-effector, *Proc. 5th Int. Conf. on Robot Vision and Sensory Control.*

Ballard, D. H. and Brown, C. M. (1982) *Computer Vision,* Prentice-Hall, New Jersey.

Batchelor, B. G., Hill, D. A. and Hodgson, D. C. (1984) *Automated Visual Inspection,* IFS (Publications) Ltd.

Chappel, A. (1976) *Optoelectronics—Theory and Practice,* Texas Instruments Ltd.

Corby, N. R. Jr. (1983) Machine vision for robotics, *IEEE Transactions on Industrial Electronics* **30,** 3.

Dillman, R. (1982) A sensor controlled gripper with tactile and non-tactile sensor environment, *Proc. 2nd Int. Conf. on Robot Vision and Sensory Control.*

Drews, P. *et al.* (1986) Optical sensor system for automated arc welding, *Robotics* **2,** 1.

Edling, G. (1986) New generation of seamtacking systems, *Proc. 9th Annual Conf. of British Robot Association.*

Electronic Automation Ltd. (1984) The EA laser ranger manual.

Fairchild (1984) Charge Coupled Devices (CCD) catalog.

Ferranti Ltd. (1982) Silicon Photocells Manual.

Hamamatsu (1983) Photomultipliers Booklet.

Hill, J. and West, G. A. W. (1982) Digital image processing—short course notes, City University.

Hollingum, J. (1984) *Machine Vision: the eyes of automation,* IFS Publications Ltd.

Honeywell-Visitronics (1984) The HDS-24 manual.

IPL (1984) The 2000 series linescan camera manual.

Ikeuchi, K. (1980) Numerical shape from shading and occluding contours in a single view, AI Memo 566, AI Lab., MIT, Cambridge, MA.

Iversen, W. R. (1983) Low cost system adds vision, *Electronics,* (November), p. 54–56.

Jarvis J. F. (1982) Research directions in industrial machine vision, *Computer,* (December), p. 55–61.

Jarvis, R. A. (1983) A laser time of flight range scanner for robotic vision, *IEEE Transactions on Pattern Analysis and Machine Intelligence.*

Kanade, T. and Somner, T. (1983) An optical proximity sensor for measuring surface position and orientation for robot manipulation, *SPIE* **449**.

Kaisto, I. *et al.* (1983) Optical range finder for 1.5–10 m distances, *Applied Optics* **22,** no. 20.

Karrer, H. E. and Dickey, A. M. (1983) Ultrasound imaging: an overview, *Hewlett–Packard Journal* **34,** no. 10.

Keller, E. L. (1983) Clever robots set to enter industry en masse, *Electronics,* (November), p 116–18.

Kulyasov, A. G. *et al.* (1973) Optical phase difference rangefinder with digital readout, *Soviet Journal of Optical Technology* **40,** no. 6.

Joyce-Loebl (1986) Choosing a video input, *Sensor Review* **6,** no. 2.

Lambert, J. (1984) Snap sight system is a snip, *Practical Robotics,* Sept/Oct.

Lewis, R. A. and Johnson, A. R. (1977) A scanning laser rangefinder for a robotic vehicle, *Proc. 5th Int. Joint Conf. on Artificial Intelligence.*

Loughlin, C. and Hudson, E. (1982) Eye-in-hand robot vision, *Proc. 2nd Int. Conf. on Robot Vision and Sensory Control.*

Marr, D. and Poggio, T. (1976) Cooperative computation of stereo disparity, *Science* **194**.

Marr, D. and Poggio, T. (1977) A theory of human stereo vision, Report no. AIM451, Artificial Intelligence Lab., MIT, Cambridge, MA.

Marr, J. (1986) Put your Apple in the picture, *Apple User,* February.

Matthys, R. J. *et al.* (1980) CO_2 TEA laser rangefinder, *SPIE* **227**.

Meta Machines Ltd. (1985) Advancing sensor technology into industry, *The Industrial Robot,* March.

Nelson, F. Jr. (1984) 3-D robot vision, *Robotics World,* (March), p. 28–29.

Nevatia, R. (1981) *Machine Perception,* Prentice-Hall, New Jersey.

Nimrod, N., Margalith, A. and Mergler, H. (1982) A laser based scanning rangefinder for robotic applications, *Proc. 2nd Int. Conf. on Robot Vision and Sensory Control.*

Nitzan, D., Brain, A. and Duda, R. (1977) The measurement and use of registered reflectance and range data in scene analysis, *Proc. of the IEEE* **65,** no. 2.

Okada, T. (1978) A short range finding sensor for manipulators, *Bull. Electrotech. Lab.* **42**(6), pp. 492–504.

Okada, T. (1982) Development of an optical distance sensor for robots, *Int. Journal of Robotics Research* **1,** no. 4.

Oldeft. Seampilot: a 3-D laser based seam tacking system for arc welding automation.

Owen, G. and Eckstein, R. (1982) Automating tedious tasks in the laboratory, *Analytical Chemistry* .

Page, C. J. and Hassan, H. (1981) Non-contact inspection of complex components using a rangefinder vision system, *Proc. 1st Int. Conf. on Robot Vision and Sensory Control.*

Page, C. J., Hassan, H. and Payne, D. B. (1983) An integrated rangefinding and intensity sensing system, *Proc. IEE Seminar on UK Robotics Research,* Digest no. 1983/104.

PERA (1981) Robots, report no. 361.

PERA (1982) Vision systems, report no. 366.

Polaroid. Ultrasonic rangefinder manual.

Popplestone, R. J. *et al.* (1975) Forming models of plane-and-cylinder faceted bodies from light stripes, *Proc. 4th Int. IJCAI.*

Pratt, W. K. (1978) Digital image processing, John Wiley & Sons, New York.

Pugh, A. (1982) Second generation robotics, *Proc. 12th Int. Symp. on Industrial Robots.*

Pugh, A. (1983) Robot vision, IFS (Publications) Ltd.

Robertson, B. E. and Walkden, A. J. (1983) Tactile sensor system for robotics, *Proc. 3rd Int. Conf. on Robot Vision and Sensory Control.*

Rooks, B. (1986) Advanced technology shines on the Rover 800 line, *The Industrial Robot* **13,** no. 3.

Rosenfeld, A. and Kak, A. C. (1976) *Digital Picture Processing,* Academic Press, New York.

Ruocco, S. R. (1986) The design of a 3-D vision sensor suitable for robot multisensory feedback, *Proc. 6th Int. Conf. on Robot Vision and Sensory Control.*

Ruocco, S. R. and Seals, R. C. (1986) A robot multisensory feedback system, *Proc. 9th Annual Conf. of the British Robot Association.*

Schneiter, J. L. and Sheridan, T. B. (1984) An optical tactile sensor for manipulators, *Robotics and Computer Integrated Manufacturing* **1,** no. 1.

Seals, R. (1984) Location of a mobile robot using landmarks, *Proc. Conf. on Manufacturing Eng.*

SILMA Inc. (1984) Technology background in machine vision.

Stauffer, N. and Wilwerding, D. (1984) *Electronic Focus for Cameras,* Honeywell-Visitronics.

Sugihara, K. (1977) Dictionary guided scene analysis based on depth information, in progress report on 3-D object recognition, Bionics research section, ETL, Tokyo.

Tanwar, L. S. (1984) An electro-optical sensor for microdisplacement measurement and control, *Journal of Physics-E* **17,** no. 10.

Van de Stadt, A. (1985) 3-D robot vision with one hand held camera, using PAPS, *Proc. 5th Int. Conf. on Robot Vision and Sensory Control.*

Van Gerwen, J. W. C. M. and Vleeskens, F. T. (1985) A sensory controlled multifunctional gripper, *Proc. 5th Int. Conf. on Robot Vision and Sensory Control.*

Woodham, R. J. (1978) Photometric stereo: a reflectance map technique for determining surface orientation from image intensity, *Proc. 22nd Int. Symp. SPIE.*

Zeuch, N. and Miller, R. K. (1986) *Machine Vision,* Prentice-Hall, New Jersey.

Index